Lecture Notes in Mathematics

Edited by A. Dold, F. Takens and B. Teissier

Editorial Policy
for the publication of monographs

1. Lecture Notes aim to report new developments in all areas of mathematics – quickly, informally and at a high level. Monograph manuscripts should be reasonably-self-contained and rounded off. Thus they may, and often will, presentnot only results of the author but also related work by other people. They may bebased on specialized lecture courses. Furthermore, the manuscripts should provide sufficient motivation, examples and applications. This clearly distinguishes Lecture Notes from journal articles or technical reports which normally are very concise. Articles intended for a journal but too long to be accepted by most journals, usually do not have this "lecture notes" character. For similar reasons it is unusual for doctoral theses to be accepted for the Lecture Notes series.

2. Manuscripts should be submitted (preferably in duplicate) either to one of the series editors or to Springer-Verlag, Heidelberg. In general, manuscripts will be sent out to 2 external referees for evaluation. If a decision cannot yet be reached on the basis of the first 2 reports, further referees may be contacted: The author will be informed of this. A final decision to publish can be made only on the basis of the complete manuscript, however a refereeing process leading to a preliminary decision can be based on a pre-final or incomplete manuscript. The strict minimum amount of material that will be considered should include a detailed outline describing the planned contents of each chapter, a bibliography and several sample chapters.
Authors should be aware that incomplete or insufficiently close to final manuscripts almost always result in longer refereeing times and nevertheless unclear referees' recommendations, making further refereeing of a final draft necessary.
Authors should also be aware that parallel submission of their manuscript to another publisher while under consideration for LNM will in general lead to immediate rejection.

3. Manuscripts should in general be submitted in English.
Final manuscripts should contain at least 100 pages of mathematical text and should include
- a table of contents;
- an informative introduction, with adequate motivation and perhaps some historical remarks: it should be accessible to a reader not intimately familiar with the topic treated;
- a subject index: as a rule this is genuinely helpful for the reader.

Lecture Notes in Mathematics 1707

Springer
Berlin
Heidelberg
New York
Barcelona
Hong Kong
London
Milan
Paris
Singapore
Tokyo

Radosław Pytlak

Numerical Methods for Optimal Control Problems with State Constraints

Springer

Author

Radosław Pytlak
Faculty of Cybernetics
Military University of Technology
ul. Kaliskiego 2
01-489 Warsaw, Poland
E-mail: rpytlak@isi.wat.waw.pl

Library of Congress Cataloging-in-Publication Data

Pytlak, R., 1956-
Numerical methods for optimal control problems with state constraints / R. Pytlak.
p. cm. -- (Lecture notes in mathematics; 1707)
Includes bibliographical references and index.
ISBN 3-540-66214-6 (alk. paper)
1. Control theory. 2. Mathematical optimization. 3. Numerical analysis. I. Title. II. Series: Lecture notes in mathematics (Springer-Verlag) ; 1707.
QA3.L28 no. 1707
[QA402.3]
510 s--dc21
[629.8'312]
99-16712
CIP

Mathematics Subject Classification (1991): 49M10, 49J15, 90C30, 65L06

ISSN 0075-8434
ISBN 3-540-66214-6 Springer-Verlag Berlin Heidelberg New York

Typesetting: Camera-ready TEX output by the author
SPIN: 10650221 41/3143-543210 - Printed on acid-free paper

Dedicated to Ela

Preface

The material presented in this book is the result of the author's work at the Interdisciplinary Centre for Process Systems Engineering at Imperial College, London. In initial stages, for the period of first nine months, the work was supported by the British Council, then directly by the Centre where the author was employed until August 1998.

Originally, the aim of the work was to provide numerical methods for optimal control problems with state constraints. At that time, that is, in early 1990s, the methods available for these problems (with the exception of the work [42], which is discussed later) were either the methods based on full discretization (parametrization of state and control variables), or they were function space algorithms (where decision variables were measurable functions). The first group of methods assumed *a priori* discretization of system equations, and thus excluded the application of variable stepsize integration procedures. The other group of methods was, in fact, theoretical work on the convergence of algorithms which have never been implemented.

The only exception mentioned earlier was the work by Fedorenko ([42]), who proposed the approach without any reference to a discretization of system equations. It turns out that his method can be easily adapted to implementations which do not have to specify integration procedures. Furthermore, the method was implemented and results of its application were both telling and encouraging. The major drawback of the method was the lack of its convergence analysis. In fact the method, as it was formulated, wasn't globally convergent.

The obvious task of my research project was to look closely at the Fedorenko's algorithm and the convergence analysis carried out in papers describing the methods from the second group mentioned above. This resulted in the formulation of a family of methods which combine some features of the Fedorenko's method and, at the same time, have global convergence properties. Furthermore, the new methods seem to have superior numerical properties and the convergence analysis accompanying algorithms significantly generalizes results published so far on the subject. The family of methods and their convergence analysis is presented in Chapter 3. Special attention is drawn to a feasible directions algorithm because it is the most promising first order algorithm within the family.

Further, second order methods are discussed in Chapter 5. Their global convergence properties stem from the fact that they are based on first order methods introduced in Chapter 3. The notable feature of the methods is that they generate superlinearly convergent sequences if controls are approximated by piecewise constant functions and some standard sufficiency conditions are met at stationary points. These methods do not require *a priori* discretization of state functions. Significantly, it is the first thus–formulated result in the literature on optimal control algorithms. The methods can be easily adapted to semi–infinite programming problems and other nonlinear programming problems with redundant constraints. Again, they would be the first superlinearly convergent algorithms for these problems. The approach is presented on an l_∞ exact penalty function algorithm. The particular choice of a second order algorithm resulted from its relative simplicity and popularity which, in turn, stems from its numerical properties discussed in Chapter 5.

During my work on numerical methods for control problems with state constraints the Centre was involved in the project on developing a simulation package for chemical processes which are described, possibly, by large–scale differential–algebraic equations. The simulation package was intended to be supplemented by an optimization program for optimal control problems. The question was whether it would be possible to construct such a method. It is worth pointing out that the existing packages for optimal control problems could cope, in reasonable computing time, only with problems of moderate size. An attempt to answer this question is made in Chapter 6 where a new approach to optimal control problems is outlined. The approach is based on an implicit Runge–Kutta integration procedure for fully implicit differential–algebraic equations. The method was proved to be both very efficient and able to cope with systems described by almost 3000 equations.

The appendix offers some remarks on numerical software developed in connection with the project. Special attention is paid to a quadratic programming procedure for direction finding subproblems. It is a range–space method for piecewise–linear quadratic problems.

Several people contributed to the book. First of all, Richard Vinter who greatly influenced Section 1 of Chapter 2 and Chapters 3–4 which are based on our joint papers [108], [109].

The efficient implementation of numerical procedures presented here would not be possible without efforts of people who proposed their optimal control problems and were courageous enough to apply new methods to their problems. The contributions of Dr Oscar Rotava and Dr Jezri Mohideen, who also performed numerical experiments with gOPT and CONOPT programs (discussed in Chapter 6), are gratefully acknowledged. Special thanks are to Dr Celeste Colantonio and Imperial Chemical Industries for their work on the problem described in Example 5 of Chapter 6 ([33]).

Several sections of the book have been already published. Chapters 2–4 are based on [104], [105], [92] and [108], [109] mentioned earlier. Chapter 5 extends the ideas presented in [98] while Chapter 6 describes the results of [100]. The appendix summarizes the author's work on a special quadratic programming procedure ([95], [96]) and material presented there is quite technical, especially §A.4.2 which is included for the completeness of the presentation—also shows the complexity of an efficient QP procedure implementation.

R. Pytlak

June, 1999
Warsaw

Contents

1 Introduction **1**
1 The Calculus of Variations 1
2 Optimal Control 5
3 Numerical Methods for Optimal Control Problems 7

2 Estimates on Solutions to Differential Equations and Their Approximations **13**
1 Linear Approximations 13
2 Lagrangian, Hamiltonian and Reduced Gradients 19

3 First Order Method **27**
1 Introduction 27
2 Representation of Functional Directional Derivatives 31
3 Relaxed Controls 32
4 The Algorithm 34
5 Convergence Properties of the Algorithm 38
6 Proof of the Convergence Theorem, etc. 41
7 Concluding Remarks 52

4 Implementation **55**
1 Implementable Algorithm 55
1.1 Second Order Correction To the Line Search 65
1.2 Resetting the Penalty Parameter 66
2 Semi–Infinite Programming Problem 66
3 Numerical Examples 68

5 Second Order Method **81**
1 Introduction 81
2 Function Space Algorithm 84
3 Semi–Infinite Programming Method 86
4 Bounding the Number of Constraints 92
4.1 Some Remarks on Direction Finding Subproblems . 94
4.2 The Nonlinear Programming Problem 98

4.3 The Watchdog Technique for Redundant Constraints 107
4.4 Two-Step Superlinear Convergence 121
4.5 Numerical Experiments 125
5 Concluding Remarks 127

6 Runge–Kutta Based Procedure for Optimal Control of Differential — Algebraic Equations 129
1 Introduction 129
2 The Method 133
2.1 Implicit Runge–Kutta Methods 134
2.2 Calculation of the Reduced Gradients 137
3 Implementation of the Implicit Runge–Kutta Method . . . 144
3.1 Simplified Newton Iterations 144
3.2 Stopping Criterion for the Newton Method 145
3.3 Stepsize Selection 146
4 Numerical Experiments 151
5 Some Remarks on Integration and Optimization Accuracies 164
6 Concluding Remarks 166

A A Primal Range–Space Method for Piecewise–Linear Quadratic Programming 169
A.1 Software Implementation 169
A.2 A Range–Space Method–Introduction 170
A.3 The Basic Method 171
A.4 Efficient Implementation 175
A.4.1 Adding a Bound to the Working Set 178
A.4.2 Deleting a Bound from the Working Set 182
A.4.3 Adding a Vector a to the Working Set 184
A.4.4 Deleting a Vector a from the Working Set 186
A.5 Computation of the Lagrange Multipliers... 187
A.6 Modifications and Extensions 188
A.7 Numerical Experiments 191

References 197

List of Symbols 209

Subject Index 213

List of Tables

4.1 Example 1, summary of results. 71
4.2 Example 2, summary of results. 72
4.3 Example 3, summary of results. 73
4.4 Parameters of the windshear problem. 76

5.1 Performance of *FD Algorithm*. 126

6.1 Runge–Kutta methods: order of convergence. 136
6.2 Example 1, CPU time (sec). 153
6.3 Example 2, CPU time (sec). 155
6.4 Example 3, CPU time (sec). 157
6.5 Example 4, CPU time (sec). 157
6.6 Performance of SQP algorithm. 161

A.1 Comparison of LSSOL and PNTSOL codes on the problems (**PLQP**) with diagonal Hessian matrices. 194
A.2 Comparison of LSSOL and PNTSOL codes on the problems (**PLQP**) with dense Hessian matrices. 195

List of Figures

4.1 Approximation of state constraint. 56
4.2 Example 1, optimal control and state trajectories for $N =$ 1000. 71
4.3 Example 2, optimal control for $N = 1000$. 74
4.4 The windshear problem, optimal state trajectories. 78
4.5 The windshear problem, optimal state trajectories. 79

5.1 Geometry of a superlinearly convergent sequence. 113

6.1 Error estimates, Example 2. 150
6.2 Error estimates, Example 3. 150
6.3 Error estimates, Example 4. 151
6.4 Example 1, control profile. 154
6.5 Example 2, CO_2 absorption/stripping process. 155
6.6 Example 2, control profiles. 156
6.7 Examples 3–4, distillation column. 158
6.8 Example 4, optimal control profiles. 159
6.9 Example 4, control profiles—sensitivity eqns. approach. . . 160
6.10 Example 5, PFD process. 162
6.11 Example 5, optimal temperature—T_{in1}, short horizon. . . . 164
6.12 Example 5, optimal output—T_{out1}, comparison of sensitivity eqns. approach with our method. 165

1

Introduction

We provide very brief historical overview of the calculus of variations and optimal control. An introduction to computational methods for optimal control problems is also presented.

1 The Calculus of Variations

The aim of the book is to describe some numerical approach to optimal control problems which, in its simplest nontrivial form, can be stated as

$$\min_{(x,u)} \left[F_0(x,u) = \int_{t_0}^{t_1} f^0(t, x(t), u(t))dt \right] \tag{1.1}$$

subject to the constraints

$$\dot{x}(t) = f(t, x(t), u(t)), \quad t \in [t_0, t_1], \tag{1.2}$$

$$x(t_0) = x_0, \quad x(t_1) = x_1. \tag{1.3}$$

Here, $t \to u(t) \in \mathcal{R}^m$ is the *control function* and $t \to x(t) \in \mathcal{R}^n$ is the *state trajectory*.

For the special case of $\dot{x} = f = u$ $(n = m)$ we can rewrite the problem (1.1)–(1.3) as follows:

$$\min_{x} \int_{t_0}^{t_1} f^0(t, x(t), \dot{x}(t))dt \tag{1.4}$$

subject to the constraints (1.3). The problem (1.4), (1.3) is in the centre of the calculus of variations. The notable feature of this problem is that the minimization takes in the space of all trajectories x.

Optimal control problems (1.1)–(1.3), by contrast, involve minimization over a set of trajectories which satisfy dynamical constraints (1.2). While all nontrivial features of the problem of the calculus of variations arise because of the functional f^0, the optimal control theory contains problems where f^0 is 1, i.e. completely trivial, and therefore all the interesting action occurs because of the dynamics f. Such problems, in which we minimize time $t_1 - t_0$, are called *minimum time problems.* It is in these problems that the difference between optimal control and the calculus of variations is most clearly seen. It is not surprising that these problems stimulated the early development of optimal control in the 1960s and it is regarded that the birth of optimal control coincides with the formulation of the first minimum time problem ([51], [116], [24]).

The minimum time problem in question is the brachistochrone problem, which according to [51], was posed by Galileo Galilei in 1638. He also proposed the incorrect conjecture on the solution. Johann Bernoulli challenged his colleagues to solve the problem in 1696; not only did he solve it himself (in 1697), but so did Leibniz, Newton (anonymously), Jacob Bernoulli (brother of Johann), Tschirnhaus and l'Hopital ([116]).

With the work of Johann and Jacob Bernoulli, Leibniz, Tschirnhaus, Newton and l'Hopital on the brachistochrone problem, the calculus of variations (and optimal control) started its progress marked by the names of prominent mathematicians.

L. Euler, who was a student of Bernoulli, published a treatise in 1744 called *The Method of Finding Plane Curves that Show Some Property of Maximum or Minimum.* J.L. Lagrange corresponded in 1755 with Euler and showed how to remove the need for geometrical insight in Euler's method and how to reduce it to a quite analytical machine. Lagrange derived the necessary condition, for x to be a minimizer of (1.4), (1.3), known today as the Euler–Lagrange equation:

$$\frac{d}{dt}\left[\frac{\partial f^0}{\partial u}(t,x(t),\dot{x}(t))\right] = \frac{\partial f^0}{\partial x}(t,x(t),\dot{x}(t)), \tag{1.5}$$

where $f^0(t,x,u)$ is the function on $\mathcal{R}^{2n+1}$.

The second variations were introduced to the calculus of variations by A.M. Legendre in 1786 who stated a new necessary condition. It says that along a minimizer of (1.4), (1.3)

$$\frac{\partial^2 f^0}{\partial u^2}(t,x(t),\dot{x}(t)) \tag{1.6}$$

is a nonnegative definite matrix. But only in 1836 K.G.J. Jacobi gave them more insight. He also showed that the partial derivatives of the performance index with respect to each state obeyed a certain partial differential equation. This equation was discovered almost at the same time by W.R. Hamilton who also introduced a new formalism to the calculus of variations.

The new formalism is based on the Hamiltonian function

$$H(t,\xi,\pi) = \mu^T\pi - f^0(t,\xi,\mu) \tag{1.7}$$

in which $\xi,\mu,\pi \in \mathcal{R}^n$ and μ is the function of (t,ξ,π) defined implicitly by the equation

$$\pi = f^0_\mu(t,\xi,\mu)^T$$

where

$$f_\mu^0(t,\xi,\mu) = \frac{\partial f^0}{\partial \mu}(t,\xi,\mu).$$

Consider now the Hamiltonian

$$H(t,x(t),p(t)) = \dot{x}(t)^T p(t) - f^0(t,x(t),\dot{x}(t)), \tag{1.8}$$

along the state trajectory x, defined by

$$\dot{x}(t) = \phi(t,x(t),p(t)), \tag{1.9}$$

where p satisfies

$$p(t) = f_u^0(t,x(t),\dot{x}(t))^T. \tag{1.10}$$

The solution of (1.10) is obtained then through the equation:

$$\dot{x}(t) = H_p(t,x(t),p(t))^T.$$

Indeed, when (1.9) holds we have

$$\begin{aligned} H_p(t,x(t),p(t)) &= \frac{\partial}{\partial p}\left[\dot{x}(t)^T p(t) - f^0(t,x(t),\dot{x}(t))\right] \\ &= \dot{x}(t)^T + p(t)^T \phi_p(t,x(t),p(t)) - \\ &\quad f_u^0(t,x(t),\dot{x}(t))\phi_p(t,x(t),p(t)) \\ &= \dot{x}(t)^T \end{aligned}$$

We also note that when $\dot{x}(t)$ is expressed by (1.9) then, from (1.8), we have

$$\begin{aligned} H_x(t,x(t),p(t)) &= p(t)^T \phi_x(t,x(t),p(t)) - f_x^0(t,x(t),\dot{x}(t)) - \\ &\quad f_u^0(t,x(t),\dot{x}(t))\phi_x(t,x(t),p(t)) \\ &= -f_x^0(t,x(t),\dot{x}(t)). \end{aligned} \tag{1.11}$$

This implies that the Euler–Lagrange equations (1.5) can be written as follows

$$\dot{x}(t) = H_p(t,x(t),p(t))^T \tag{1.12}$$

$$\dot{p}(t) = -H_x(t,x(t),p(t))^T. \tag{1.13}$$

Consider now the trajectory x connecting (t_0,x_0) and (t,ξ). Assume that x has been perturbed by δx and that $x+\delta x$ passes through (t_0,x_0) and $(t+\delta t,\xi+\delta\xi)$. We can show, as in [45], that

$$\delta F_0 = \int_{t_0}^{t+\delta t} f^0(\tau,x(\tau)+\delta x(\tau),\dot{x}(\tau)+\delta\dot{x}(\tau))d\tau -$$

$$\int_{t_0}^{t} f^0(\tau, x(\tau), \dot{x}(\tau))d\tau =$$
$$\int_{t_0}^{t} \left[f_x^0(\tau, x(\tau), \dot{x}(\tau)) - \frac{d}{dt} f_u^0(\tau, x(\tau), \dot{x}(\tau)) \right] \delta x(\tau) d\tau +$$
$$f_u^0(t, \xi, \dot{x}(t))\delta\xi + \left[f^0(t, \xi, \dot{x}(t)) - f_u^0(t, \xi, \dot{x}(t))\dot{x}(t) \right] \delta t. \tag{1.14}$$

(We derive these conditions by using Taylor's theorem and then the differentiation by part formula.) If x is a minimizer of the problem (1.4), (1.3) with the endpoint condition: $x(t) = \xi$ then (1.14) reduces to

$$\delta F_0 = f_u^0(t, \xi, \dot{x}(t))\delta\xi + \left[f^0(t, \xi, \dot{x}(t)) - f_u^0(t, \xi, \dot{x}(t))\dot{x}(t) \right] \delta t \tag{1.15}$$

since x satisfies (1.5).

If we consider the perturbation in the variables t, x, p, H defined by (1.8)–(1.10) then δF_0 in (1.14) can be expressed as follows

$$\delta F_0 = \int_{t_0}^{t} \left[f_x^0(\tau, x(\tau), \phi(\tau, x(\tau), p(\tau))) - \dot{p}(\tau)^T \right] \delta x(\tau) d\tau +$$
$$p(t)^T \delta\xi - H(t, \xi, p(t))\delta t \tag{1.16}$$

which simplifies to

$$\delta F_0 = p(t)^T \delta\xi - H(t, \xi, p(t))\delta t, \tag{1.17}$$

when x is a minimizer, since p satisfies (1.13) and (1.11) holds.

Denote now by $S(t, \xi)$ optimal values of (1.4) with varying endpoints (t, ξ):

$$S(t, \xi) = \min_x \left\{ \int_{t_0}^{t} f^0(t, x(t), \dot{x}(t))dt : \quad x(t_0) = x_0, \ x(t) = \xi \right\}. \tag{1.18}$$

We will show that the function S plays an important role in the calculus of variations and optimal control.

Consider

$$\delta S = S(t + \delta t, \xi + \delta\xi) - S(t, \xi) \tag{1.19}$$

and notice that value $S(t, \xi)$ is determined by the trajectory $\bar{x}$ which is a minimizer of (1.4) with the endpoint conditions: $x(t_0) = x_0$, $x(t) = \xi$. Therefore, we can use (1.17) to state

$$\delta S(t, \xi) = \delta F_0 = p(t)^T \delta\xi - H(t, \xi, p(t))\delta t. \tag{1.20}$$

We conclude that this perturbation of S will be the total differential dS in variables (t, ξ) if we have

$$p(t) = S_\xi(t, \xi), \quad -H(t, \xi, p(t)) = S_t(t, \xi).$$

We can eliminate p from these equations to get

$$S_t(t,\xi) + H(t,\xi,S_\xi(t,\xi)) = 0 \tag{1.21}$$

which is the Hamilton–Jacobi equation.

2 Optimal Control

After this very brief overview of the calculus of variations (presented with many simplifying assumptions not stated explicitly) we can conjecture necessary conditions for the basic problem (1.1)–(1.3).

Consider first system equations $\dot{x}(t) = u(t)$ and the Hamiltonian

$$H(t,x,u,p) = u^T p - f^0(t,x,u). \tag{2.1}$$

In addition to equations (1.12)–(1.13) (valid also for H defined in this way), we have

$$0 = H_u(t,x(t),u(t),p(t)) \tag{2.2}$$

since

$$H_u(t,x(t),u(t),p(t)) = p(t)^T - f_u^0(t,x(t),\dot{x}(t)) = 0 \tag{2.3}$$

from (1.10).

In order to deal with problem (1.1)–(1.3) we have to modify the Hamiltonian further by introducing the 'abnormal multiplier' p_0 corresponding to the objective index f^0 and by replacing u with f:

$$H(t,x,u,p,p_0) = p^T f(t,x,u) - p_0 f^0(t,x,u). \tag{2.4}$$

Conditions (1.12)–(1.13), (2.2) can be extended to problem (1.1)–(1.3) by using the Hamiltonian (2.4):

$$\begin{aligned} \dot{x}(t) &= H_p(t,x(t),u(t),p(t),p_0)^T &(2.5)\\ \dot{p}(t) &= -H_x(t,x(t),u(t),p(t),p_0)^T &(2.6)\\ 0 &= H_u(t,x(t),u(t),p(t),p_0), &(2.7) \end{aligned}$$

where $(\int_{t_0}^{t_1} |p(t)|dt, p_0) \neq (0,0)$—nontriviality condition. p is now postulated as a function which together with x and u satisfy the above equations—equation (2.7) replaces equation (1.10) as (2.3) suggests. Furthermore, the condition (2.7) can be strengthened to[1]

$$H(t, x(t), u(t), p(t), p_0) = \max_{\tilde{u}} H(t, x(t), \tilde{u}, p(t), p_0). \tag{2.8}$$

The Hamilton-Jacobi equation (1.21) can also be extended to optimal control problems (1.1)–(1.3). Denote by $V(t, \xi)$ the optimal value function,

$$V(t, \xi) = \min_{v} \left\{ \int_t^{t_1} f^0(t, x(t), v(t))dt : \dot{x}(t) = f(t, x(t), v(t)), \right.$$
$$\left. x(t) = \xi, \ x(t_1) = x_1 \right\},$$

and by $v(\tau, t, \xi)$, $\tau \in [t, t_1]$ the corresponding optimal control. If $u(t, \xi) = v(t, t, \xi)$ then the following equation holds (see, e.g., [11]):

$$V_t(t, \xi) + H(t, \xi, u(t, \xi), V_\xi(t, \xi), -1) = 0. \tag{2.9}$$

which, if we substitute $S(t, \xi) = -V(t, \xi)$, can also be stated as

$$S_t(t, \xi) + H(t, \xi, u(t, \xi), S_\xi(t, \xi), 1) = 0.$$

The important difference between problems of the calculus of variations and those of optimal control is that the latter can include various types of constraints, among them control constraints of the form $u(t) \in \Omega$ where Ω is some convex compact set in $\mathcal{R}^m$. In this case we can show that (2.5)–(2.7) are still valid with condition (2.7) replaced by

$$H(t, x(t), u(t), p(t), p_0) = \max_{\tilde{u} \in \Omega} H(t, x(t), \tilde{u}, p(t), p_0). \tag{2.10}$$

Equations (2.5)–(2.6), (2.10), or their variants, play an important role in optimal control computations. They constitute the Pontryagin's Maximum Principle established in [87].

Furthermore, the Hamilton–Jacobi equation (2.9) can be stated as the so–called Bellman's dynamic programming equation:

$$V_t(t, \xi) = \max_{u \in \Omega} \left[-H(t, \xi, u, V_\xi(t, \xi), -1) \right]. \tag{2.11}$$

Other constraints can be imposed in optimal control problems such as *state constraints* (also called pathwise inequality constraints):

$$q(t, x(t)) \le 0 \ \forall t \in [t_0, t_1]. \tag{2.12}$$

---

[1]Notice that if matrix (1.6) is positive definite and (2.2) holds then the Legendre condition is a local case of (2.8). This follows from the fact that problem (2.1), (1.3) can be stated as an optimal control problem with $\dot{x} = f = u$ for which the Hamiltonian (2.1) applies.

The Weierstrass (1815–1897) 'global' condition ([45],[128]) is even more closely related to condition (2.8). For the lack of space Weierstrass' revolutionary contribution to the calculus of variations is not discussed here.

and conditions (2.5)–(2.6), (2.9) can be extended to control problems with state constraints. However, we do not state these rather complicated equations here (equations (2.5)–(2.6), (2.8) for this case are derived in Chapter 3) since the aim of the introduction is to give very short overview of computational methods for optimal control problems. The basic necessary conditions presented in this chapter are given in support of this overview.

3 Numerical Methods for Optimal Control Problems

Computational methods based on the Bellman's equation (2.11) were among the first proposed for optimal control problems ([9], [10]). These methods are especially attractive when an optimal control problem is discretized *a priori* and the discrete version of equation (2.11) is used to solve it. This significantly reduces the computer memory requirement for storing values of $u(t, \xi)$[2] although even in this case the method can hardly be applied to problems with more than a few states. Some other approximation schemes (see, e.g., [13]) have been proposed to cope with the 'curse of dimensionality' (Bellman's own phrase) with rather limited success in the author's opinion (shared with that of Bryson in [24]). This is mainly due to the fact that the solution to the Bellman's equation gives also *closed-loop* solution in the form of the function $u(t, \xi)$ while other computational methods discussed here guarantee only *open-loop* solutions u as functions of time only. Nonetheless, dynamic programming algorithms are widely used by scientists, engineers and economists ([13], [71]).

Some of the first numerical methods for optimal control problems were based on the *shooting method* of guessing initial values for $p(t_0)$, integrating equations (2.5)–(2.6), and then adjusting $p(t_0)$ to meet some terminal conditions imposed on $p(t_1)$ (these conditions are stated in the subsequent chapters). Controls are forced to satisfy (2.7) ([20]). This method, which was extended to problems with state constraints (2.12) ([81], [27], [28]), usually guarantees very accurate solutions provided that good initial guesses for p are available. The shooting method is further discussed in Chapter 4.

Gradient algorithms were proposed by Kelley ([6]) and by Bryson and Denham ([25]). These methods evaluate gradients for the functional F_0 by integrating forward systems equations (2.5) to obtain the state trajectory x. This trajectory is then used to evaluate equations (2.6), backward in time. x and p are used to calculate the gradient of F_0 but with respect to control variables u only. This is possible due to the fact that x is uniquely

[2] We solve equation (2.11) backward in time. $u(t_0, x_0)$ is calculated at initial time and then the optimal control is determined by evaulating system equations using $u(t, \xi)$.

determined by u, thus F_0 can be considered as a function of u only. The exact formula for this gradient is given in §2.2 (c.f., (2.2.20)–(2.2.24)).

There are many variants of gradient algorithms depending on whether the problem is *a priori* discretized in time, and on the optimization solver used. There are also second order gradient algorithms which are related to *accessory problems* ([26]).[3] However, these require substantially more programming than first order methods and require analytic expressions for the second order derivatives ([21]). The calculation of second order derivatives can be avoided (at the expense of a slower rate of convergence of the optimization algorithm) by applying an SQP (Sequential Quadratic Programming) type optimization algorithm in which these derivatives are approximated by quasi–Newton formulas ([43], [12]).

The gradient algorithms can also be applied to optimal control problems with state constraints (2.12) if the problem is discretized *a priori* and the discretization for states coincides with that for controls. If the second condition does not hold then the optimal control problem can have *redundant constraints* and as a result the rate of convergence of second order methods can deteriorate. This issue is further discussed in Chapter 5.

Collocation methods are similar to the gradient algorithms applied to *a priori* discretized problems with the exception that gradients are not calculated with the help of equations (2.6). Instead both x and u are regarded as optimization parameters and equations (2.5) are considered as nonlinear equality constraints in the discretized problem. Any general NLP (NonLinear Programming) algorithm can be applied to the problem but its *reduced gradient* version (or *generalized reduced gradient* version) ([1], [38], see also [80]) is the most efficient in this application. When state constraints (2.12) are simple 'box' constraints then they can be easily dealt with by the collocation methods (this was one of the reasons for their development—[117], [15], [127]). The collocation methods are also discussed in Chapter 6.

The gradient algorithms and the collocation methods are widely used and are regarded as the most reliable, if not very accurate, methods for optimal control problems with state constraints ([24]). The main reason for their reliability is that they do not require the knowledge of the number of *active arcs* of state constraints ([81], [27], [57]). If this knowledge is available, together with their approximate locations on the time horizon, then the shooting methods provide solutions with much higher accuracy. However, the analytical pre–analysis required by shooting methods is very difficult (if possible at all) for problems with more than a few states. On the other hand, the gradient algorithms solutions are often sufficiently accurate for engineering purposes ([24]).

---

[3] Accessory problems are linear–quadratic time varying problems, i.e., problems with linear dynamics and quadratic performance indices whose weighting matrices are the second derivatives of the Hamiltonian with respect to the states and the controls.

There are two major sources of the inaccuracies in the gradient algorithms: 1) errors caused by integration schemes applied to system equations (1.2) and the objective index (1.1); 2) errors introduced by the optimization procedure.

Approximation errors. In order to solve an optimal control problem by the gradient algorithm the first thing we have to decide is in what class of functions we look for an optimal solution. The most widely used class is that of piecewise constant functions (piecewise linear or piecewise polynomial functions are also applied). Once the class of control functions has been chosen we have to decide on the integration procedure for the system equations. The choice of the integration procedure can have significant effect on reducing the approximation errors. The proper integration procedure with small steps can almost eliminate the approximation errors even without upgrading the class of functions used for control approximations.

Optimization errors. These errors stem from the fact that the optimization procedure usually performs only a finite number of iterations. They can be reduced by performing more iterations, or by applying the second order optimization procedure which guarantees a higher rate of convergence (but whose cost of an iteration is much higher than that of a first order method). Since any number is represented in a computer memory with a finite precision, additional errors are introduced and cannot be eliminated without using computers which guarantee lower *roundoff errors.*

To illustrate the approximation errors consider the brachistochrone problem which was already mentioned. Two points $P_0 = (t_0, x_0)$ and $P_1 = (t_1, x_1)$ are given in a vertical plane with P_0 higher than P_1. Point mass acted upon solely by gravitational forces is to move along a trajectory C joining P_0 and P_1. The problem is to find the trajectory C so that the time required for the mass to go from P_0 to P_1 is a minimum. The trajectory which solves the problem is called the brachistochrone and it is a cycloid—the trajectory described by a point P in a circle that rolls without slipping on the t axis, in such a way that P passes through P_0 and then P_1, without hitting the t axis in between (c.f. Figure 4 in [116]).

It can be shown that the problem of finding the optimal trajectory C can be stated as the problem of the calculus of variations ([11]):

$$\min_x \left[F_0(x, \dot{x}) = \int_{t_0}^{t_1} \sqrt{\frac{1 + \dot{x}(t)^2}{x(t) - x_0}} dt \right] \tag{3.1}$$

subject to the constraints

$$x(t_0) = x_0, \quad x(t_1) = x_1. \tag{3.2}$$

Problem (3.1)–(3.2) can be also formulated as an optimal control problem:

$$\min_u \left[F_0(x, u) = \int_{t_0}^{t_1} \sqrt{\frac{1 + u(t)^2}{x(t) - x_0}} dt \right] \tag{3.3}$$

subject to constraints (3.2) and

$$\dot{x}(t) = u(t). \tag{3.4}$$

Following [42] consider the brachistochrone problem with, $t_0 = 0$, $t_1 = 2$, $x_0 = 0$ and $x_1 = 1.2$. It can be shown that the minimum value in (3.1), for this particular problem, is ≈ 3.547.

Assume that x is approximated by piecewise linear functions. To this end we divide $[t_0, t_1]$ ($t_0 = 0$, $t_1 = 2$) into $N = 100$ equal subintervals:

$$t_0 = \tau_0 < \tau_1 < \tau_2 < \ldots < \tau_N = t_1; \ \tau_k = k\tau, \ \tau = (t_1 - t_0)/N. \tag{3.5}$$

The approximation of x is defined by $x_k = x(\tau_k)$, between the mesh points we assume that x is linear. If we denote by x^N the function defined in this way, then the integral in (3.1) can be approximated by the formula

$$F_0(x^N, \dot{x}^N) = \tau \sum_{k=0}^{N-1} \sqrt{\left[1 + \frac{(x_{k+1} - x_k)^2}{\tau^2}\right] / \frac{(x_k + x_{k+1})}{2}}. \tag{3.6}$$

Consider now the trajectory defined by the points x_k:

$$x_0 = 0, \ x_1 = x_2 = x_3 \ldots = x_N = 1.2. \tag{3.7}$$

Using (3.6) we can evaluate the integral in (3.1) as

$$\begin{aligned} F_0(x^N, \dot{x}^N) = \tau\sqrt{\frac{1 + (1.2/\tau)^2}{1.2/2}} + (N-1)\tau\sqrt{1/1.2} &\approx 1.549 + 1.807 \\ &= 3.356 \end{aligned}$$

which is much lower value than the optimal value ≈ 3.547. Surely this suspiciously low value must be due to the simplified formula (3.6).

Knowing that we have to improve the integrating formula for F_0 consider optimal control problem (3.2)–(3.4). We assume that controls are piecewise constant

$$u(t) = u_k \quad \text{for} \ \ \tau_k \le t < \tau_{k+1}, \ k = 0, \ldots, N-1$$

and τ_k are defined by (3.5). We denote this control by u^N. This time the integral in (3.3) is evaluated as follows

$$F_0(x^N, u^N) = \sum_{k=0}^{N-1} \int_{\tau_k}^{\tau_{k+1}} \sqrt{\frac{1 + u_k^2}{x_k + (t - \tau_k)u_k}} dt \tag{3.8}$$

and can be expressed as

$$F_0(x^N, u^N) = \sum_{k=0}^{N-1} \begin{cases} \tau\sqrt{2\dfrac{1 + u_k^2}{x_k + x_{k+1}}} & \text{if} \ \ |u_k| < \varepsilon \approx 0 \\ 2\dfrac{\sqrt{1 + u_k^2}}{u_k}(\sqrt{x_{k+1}} - \sqrt{x_k}) & \text{if} \ \ |u_k| \ge \varepsilon \end{cases}$$

This formula, for u^N corresponding to (3.7), gives value $\approx 2.191 + 1.807 = 3.998$.

A critical reader would complain that the brachistochrone numerical example is questionable since at time $t_0 = 0$ we have $x(0) = 0$ and as a result the singularity in the integral function of F_0. However, the example makes the point: the approximation errors can be greatly reduced by increasing the accuracy in the integration procedure even without changing the number of mesh points in piecewise constant approximations of controls. The example also shows that a higher accuracy is required at time intervals where states change rapidly. Notice that the difference in the values of F_0 given by (3.6) and (3.8) formulas is caused by the difference in their values solely on the subinterval $[0, 02]$ where $u = 60$ (1.549 for (3.6) and 2.191 for (3.8)). It is not surprising that the algorithm considered in [42] and based on formula (3.8) found an approximate solution with the value of F_0 equal to 3.553. The example suggests that using a variable stepsize integration procedure would be very desirable in the gradient algorithms.

The approach advocated in this monograph assumes that system equations are integrated by the procedure with a variable stepsize. The price we pay for reducing the approximation errors is nontrivial convergence analysis of the gradient algorithms when applied to problems with state constraints (see Chapter 5). Notice also that there are strong arguments for using variable stepsize procedures for integrating differential–algebraic equations besides the fact that these procedures are recommended for *stiff equations* (c.f. [55]). This issue is discussed in §6.7.

Ideally we would like to solve a sequence of the discretized control problems (corresponding to the continuous time problem) with the increased number of mesh points both for the approximations of control and state functions. It can be shown that the cost of the system equations evaluation is proportional to the number of mesh points in the approximation of state functions. The same estimate applies also to the evaluation of gradients (c.f. §2.2). However, the number of gradients, which the gradient algorithms have to evaluate, can be proportional to N (N is the number of mesh points in the approximation of control functions), if the state constraints have active arcs (see Chapter 5). This implies that the cost of one iteration of the gradient algorithm can be proportional to N^2 (notice that we take into account only the cost of the evaluation of gradients). The exception to this rule is the first order algorithm presented in Chapters 3–4.

The question is whether we will be able to construct computational methods for optimal control problems with state constraints such that the cost of their application is proportional to N. The gradient algorithms do not have this property; neither the collocation methods described in [117] and [15] (see also [16]). The collocation method proposed in [126] is based on an interior point quadratic programming procedure. The cost of one iteration of this algorithm is proportional to N. In practice the iteration count of the method is independent of N, though formal analyses suggest that it should be $O(N^{1/2})$ (c.f. [127]). However, the method, when applied to control problems with many states, can be costly and, in our opinion, it couldn't be efficient when applied to some problems reported in Chapter 6. This happens because the cost of its iteration is estimated by LN where L is the number which rapidly grows with the number of states.

Interesting algorithms for accessory problems are discussed in [111], [132] and [131]. The algorithms solve these problems by finding saddle points of a Lagrangian which has the property that it is decomposable with respect to state and control variables when the Lagrange multipliers are fixed, and vice versa (c.f. §2.2). Primal–dual steepest descent and conjugate gradient algorithms which exploit the structure are introduced in [132]. The cost of one iteration of both algorithms is $O(N)$ and the conjugate gradient algorithm finds a solution in a finite number of iterations. However, it has to be shown that these algorithms could be the basis of efficient algorithms for nonlinear optimal control problems.

The gradient algorithms, we propose in the book, reduce the approximation errors by increasing the accuracy of an integration procedure (by increasing the number of mesh points in the approximation of state functions). They can be applied to problems with many state variables. The *search direction* procedures of these algorithms are based on a bounded number of gradients irrespective of the integration accuracy. This is the most desired property of the gradient algorithms designed for optimal control problems with state constraints. Their convergence analysis is based on the assumption that an integration procedure does not introduce any errors. This allows us to focus on main features of these methods without resorting to elaborate (and rather academic in our opinion) analysis of the limit properties of solutions to the discretized problems. Practical issues related to the integration errors are discussed in Chapter 6.

2

Estimates on Solutions to Differential Equations and Their Approximations

Estimates on solutions to perturbed ordinary differential equations are established here.

1 Linear Approximations

In this self-contained chapter we gather together results on the relationship between nonlinear control systems and their approximations which are useful in the analysis of the convergence properties of optimal control algorithms. The control system is described by

$$\begin{aligned} \dot{x}(t) &= f(t, x(t), u(t)) \text{ a.e. on } [0,1] = T \\ x(0) &= x_0. \end{aligned} \tag{1.1}$$

Here, $f : T \times \mathcal{R}^n \times \mathcal{R}^m \to \mathcal{R}^n$ is a given function and $x_0 \in \mathcal{R}^n$ a given vector. A *control function* $u : T \to \mathcal{R}^m$ is a measurable function which satisfies $u(t) \in \Omega$ a.e. on T, where $\Omega \subset \mathcal{R}^m$ is a given subset. Control functions are thus drawn from the set

$$\mathcal{U} = \left\{ v \in \mathcal{L}^1_m[T] : \; v(t) \in \Omega \text{ a.e. on } T \right\}.$$

Given any control function the solution to (1.1) is an absolutely continuous function $x : T \to \mathcal{R}^n$ (c.f. *Proposition 1.1* below). It is denoted by x^u and is referred to as the *state trajectory corresponding to* u.

The corresponding linearized control system of interest here is

$$\begin{aligned} \dot{y}(t) &= f_x(t, x^u(t), u(t))y(t) + f_u(t, x^u(t), u(t))d(t) \\ y(0) &= 0. \end{aligned} \tag{1.2}$$

The solution to (1.2), which depends on $u \in \mathcal{U}$ and $d \in \mathcal{L}^1_m[T]$, is written $y^{u,d}$.

We refer to the hypothesis

(H1) Ω is a bounded set. $f(t, \cdot, \cdot)$ is differentiable, f, f_x and f_u are continuous and there exists $K < \infty$ such that

$$\|f_x(t, x, u)\| \le K \text{ for all } (t, x, u) \in T \times \mathcal{R}^n \times \Omega. \tag{1.3}$$

(1.3) and the fact that Ω is bounded guarantee that $||x^u|| \leq M$ for some M and that some other other estimates, expressed in *Propositions 1.1-1.3*, for system (1.1) and its linearization are satisfied. We could obtain the same estimates if (1.3) is replaced by: $||f(t,x,u)|| \leq C$ for some C and all $(t,x,u) \in T \times \mathcal{R}^n \times \Omega$.

Repeated use will be made of a basic existence/uniqueness theorem obtained, for example, by specializing Theorem 3.1.6 in [32] to differential equations and applying Gronwall's lemma (Lemma 4.2, p. 139 in [11]) to establish 'uniqueness'.

Take a function $\varphi : T \times \mathcal{R}^n \rightarrow \mathcal{R}^n$ such that $\varphi(\cdot, x)$ is measurable and $\varphi(t, \cdot)$ is continuously differentiable. Assume that there exists $k_\varphi < \infty$ such that $||\varphi_x(t,x)|| \leq k_\varphi$ for all $(t,x) \in T \times \mathcal{R}^n$. Then, if there is an absolutely continuous function z such that $\int_T ||\dot{z}(t) - \varphi(t, z(t))||dt < \infty$, there is a unique absolutely continuous function $\tilde{z}$ such that

$$\begin{aligned} \dot{\tilde{z}}(t) &= \varphi(t, \tilde{z}(t)) \text{ a.e. on } T \\ \tilde{z}(0) &= z(0) \end{aligned}$$

and

$$||\tilde{z} - z||_{\mathcal{L}^\infty} \leq e^{k_\varphi} \int_T ||\dot{z}(t) - \varphi(t, z(t))||dt.$$

Proposition 1.1 *Assume* **(H1)**. *For each $u \in \mathcal{U}$ and $d \in \mathcal{L}^1_m[T]$ (1.1) and (1.2) have unique solutions (in the class of absolutely continuous vector valued functions on T), x^u and $y^{u,d}$ respectively. Furthermore there exist finite constants c_1, c_2 and c_3 such that*

$$\begin{aligned} ||x^u||_{\mathcal{L}^\infty} &\leq c_1, \\ ||x^u - x^v||_{\mathcal{L}^\infty} &\leq c_2 \, ||u - v||_{\mathcal{L}^2}, \\ \left\|y^{u,d}\right\|_{\mathcal{L}^\infty} &\leq c_3 \, ||d||_{\mathcal{L}^1} \end{aligned}$$

for all $u, \ v \in \mathcal{U}, \ d \in \mathcal{L}^1_m[T]$. In particular

$$\left\|y^{u,d}\right\|_{\mathcal{L}^\infty} \leq c_3 \, ||d||_{\mathcal{L}^\infty}$$

if $u \in \mathcal{U}, \ d \in \mathcal{L}^\infty_m[T]$.

Proof.

Take any $u \in \mathcal{U}$. Apply the basic theorem above with $z \equiv x_0$ and $\varphi(t,x) = f(t,x,u(t))$. By **(H1)**, $\int_0^1 ||\dot{z}(t) - \varphi(t,z(t))||dt \leq \int_0^1 ||f(t,x_0,u(t))||dt \leq k < \infty$ in which the constant $k = \max\{||f(t,x_0,v)|| \ : \ v \in \Omega, \ t \in T\}$. So there exists a unique solution x^u to (1.1), satisfying $||x^u||_{\mathcal{L}^\infty} \leq c_1$ in which $c_1 = e^K k$, a constant which does not depend on our choice of u. The other two estimates are also obtained from the basic theorem. (For the first estimate set $z = x^u$ and $\varphi(t,x) = f(t,x,v(t))$ and for the second set $z \equiv 0$, $\varphi(t,y) = f_x(t,x^u(t),u(t))y + f_u(t,x^u(t),u(t))d(t)$.) ■

Proposition 1.2 *Assume* **(H1)**. *Take sequences* $\{u_i\}$ *in* $\mathcal{U}$ *and* $\{d_i\}$ *in* $\mathcal{L}_m^\infty[T]$ *such that* $\|d_i\|_{\mathcal{L}^\infty} \leq M$ *for some* $M < \infty$ *independent of* i. *Then if* $u_i \to u$ *and* $d_i \to d$ *in* $\mathcal{L}_m^1[T]$ *we have* $y^{u_i,d_i} \to y^{u,d}$ *in* $\mathcal{L}_n^\infty[T]$, *as* $i \to \infty$.

Proof.

Replace $\{u_i\}$, $\{d_i\}$ by arbitrary subsequences. We shall show that along a particular subsequence $y^{u_i,d_i} \to y^{u,d}$ in $\mathcal{L}_n^\infty[T]$. Since the limit is independent of the original subsequences chosen, this will imply that the original sequence $\{y^{u_i,d_i}\}$ converges to $y^{u,d}$ in $\mathcal{L}_n^\infty[T]$ as $i \to \infty$.

Write y_i for y^{u_i,d_i} and x_i for x^{u_i}. The sequences $\{y_i\}$ and $\{x_i\}$ are bounded in $\mathcal{L}_n^\infty[T]$ by *Proposition 1.1*. For each i we have

$$\dot{y}_i(t) = f_x(t, x_i(t), u_i(t))y_i(t) + f_u(t, x_i(t), u_i(t))d_i(t) \quad \text{a.e. on } T$$

from which it now follows that $\{\dot{y}_i\}$ is bounded in $\mathcal{L}_n^\infty[T]$. By Ascoli's Theorem ([62], p. 27), subsequences can be chosen such that $y_i \to y$ in $\mathcal{L}_m^\infty[T]$ for some Lipschitz continuous function y. Since $u_i \to u$ and $d_i \to d$ in $\mathcal{L}_m^1[T]$ we can arrange by further subsequence extraction that $u_i \to u$ and $d_i \to d$ a.e..

However for each i

$$y_i(t) = \int_0^t [f_x(s, x_i(s), u_i(s))y_i(s) + f_u(s, x_i(s), u_i(s))d_i(s)]\, ds$$

for all $t \in T$. For each $t \in T$ we may pass to the limit on both sides of this equation with the help of the dominated convergence theorem ([63], p. 121) and thereby obtain

$$y(t) = \int_0^t [f_x(s, x(s), u(s))y(s) + f_u(s, x(s), u(s))d(s)]\, ds.$$

Since this is true for all t

$$\begin{aligned} \dot{y}(t) &= f_x(t, x(t), u(t))y(t) + f_u(t, x(t), u(t))d(t) \quad \text{a.e. on } T \\ y(0) &= 0. \end{aligned}$$

But this differential equation for y has a unique solution, namely $y^{u,d}$. We have shown, for the ultimate subsequence $y^{u_i,d_i} \to y^{u,d}$ in $\mathcal{L}_n^\infty[T]$. This is what needed to be proved. ■

Proposition 1.3 *Assume* **(H1)**. *Then there exists a function* $o_1(\cdot) : (0, \infty) \to (0, \infty)$ *such that* $s^{-1}o_1(s) \to 0$ *as* $s \downarrow 0$ *and*

$$\|x^{u+d} - (x^u + y^{u,d})\|_{\mathcal{L}^\infty} \leq o_1(\|d\|_{\mathcal{L}^\infty})$$

for all $u \in \mathcal{U}$ *and* $d \in \mathcal{L}_m^\infty[T]$.

Proof.

Choose any $u \in \mathcal{U}$ and $d \in \mathcal{L}_m^\infty[T]$. Apply the basic theorem with 'comparison' function $z = x^u + y^{u,d}$ and $\varphi(t,x) = f(t,x,u(t)+d(t))$. We obtain

$$\begin{aligned}\left\|x^{u+d} - (x^u + y^{u,d})\right\|_{\mathcal{L}^\infty} \leq e^K \int_T \|\dot{x}^u(t) + \dot{y}^{u,d}(t) \\ -f(t, x^u(t) + y^{u,d}(t), u(t) + d(t))\| \, dt = \\ e^K \int_T \|f_{(x,u)}(t, x^u(t), u(t))(y^{u,d}(t), d(t)) - \\ [f(t, x^u(t) + y^{u,d}(t), u(t) + d(t)) - f(t, x^u(t), u(t))]\| \, dt \end{aligned} \qquad (1.4)$$

($f_{(x,u)}$ denotes (f_x, f_u) and $(y^{u,d}(t), d(t))$ is regarded as a column vector).

But a component-wise application of the Mean Value Theorem tells us that, for $i = 1, \ldots, n$ and $t \in T$,

$$\begin{aligned} f_i(t, x^u(t) + y^{u,d}(t), u(t) + d(t)) - f_i(t, x^u(t), u(t)) = \\ (f_i)_{(x,u)}(t, x^u(t) + \xi_i(t), u(t) + \nu_i(t))(y^{u,d}(t), d(t)) \end{aligned}$$

for some $\xi_i(t)$, $\nu_i(t)$ satisfying $\|\xi_i(t)\| \leq \|y^{u,d}(t)\|$ ($\leq c_3\|d\|_{\mathcal{L}^\infty}$ by *Proposition 1.1*) and $\|\nu_i(t)\| \leq \|d(t)\|$. We now deduce from uniform continuity of $f_{(x,u)}$ on bounded sets, *Proposition 1.1* and the boundedness of Ω that there exists $\theta : (0,\infty) \to (0,\infty)$ such that $s^{-1}\theta(s) \to 0$ as $s \downarrow 0$, and

$$\begin{aligned}\|f_{(x,u)}(t, x^u(t), u(t))(y^{u,d}(t), d(t)) - \\ [f(t, x^u(t) + y^{u,d}(t), u(t) + d(t)) - f(t, x^u(t), u(t))]\| \\ \leq \theta(\|d\|_{\mathcal{L}^\infty})\end{aligned}$$

for all $t \in T$.

It follows from (1.4) that

$$\left\|x^{u+d} - (x^u + y^{u,d})\right\|_{\mathcal{L}^\infty} \leq o_1(\|d\|_{\mathcal{L}^\infty})$$

in which $o_1(s) = e^K \theta(s)$. ∎

Proposition 1.4 *Assume* **(H1)**. *Take a function* $q : T \times \mathcal{R}^n \to \mathcal{R}$ *such that* $q(t,\cdot)$ *is differentiable and* q, q_x *are continuous. Then for any* $\varepsilon > 0$ *there exists* $o_2^\varepsilon(\cdot) : (0,\infty) \to (0,\infty)$ *such that* $s^{-1}o_2^\varepsilon(s) \to 0$ *as* $s \downarrow 0$ *and*

$$\begin{aligned}\left|\max_{t \in R_{\varepsilon,u}} [q(t, x^u(t)) + q_x(t, x^u(t))y^{u,d}(t)] - \max_{t \in T} q(t, x^{u+d}(t))\right| \\ \leq o_2^\varepsilon(\|d\|_{\mathcal{L}^\infty})\end{aligned} \qquad (1.5)$$

for all $u \in \mathcal{U}$ *and* $d \in \mathcal{L}_m^\infty[T]$, *where*

$$R_{\varepsilon,u} = \left\{ t \in T : \; q(t, x^u(t)) \geq \max_{\tilde{t} \in T} q(\tilde{t}, x^u(\tilde{t})) - \varepsilon \right\}. \qquad (1.6)$$

Proof.

Fix $\varepsilon > 0$. Define

$$\Delta q^{u,d}(t) = q(t, x^u(t)) + q_x(t, x^u(t))y^{u,d}(t).$$

By the uniform continuity of q and q_x on bounded sets, there exists $\delta_\varepsilon > 0$ such that

$$\max_{t\in R_{\varepsilon,u}} \Delta q^{u,d}(t) = \max_{t\in T} \Delta q^{u,d}(t)$$

for all $u \in \mathcal{U}$ and $d \in \mathcal{L}_m^\infty[T]$ satisfying $\|d\|_{\mathcal{L}^\infty} \leq \delta_\varepsilon$.

Now the continuous function $\Delta q^{u,d}(\cdot)$ achieves its maximum over T at some $\bar{t}$. So,

$$\max_{t\in T} \Delta q^{u,d}(t) - \max_{t\in T} q(t, x^{u+d}(t)) \leq \Delta q^{u,d}(\bar{t}) - q(\bar{t}, x^{u+d}(\bar{t})) =$$
$$q(\bar{t}, x^u(\bar{t})) + q_x(\bar{t}, x^u(\bar{t}))y^{u,d}(\bar{t}) - q(\bar{t}, x^{u+d}(\bar{t})) \leq$$
$$q_x(\bar{t}, x^u(\bar{t}))(x^{u+d}(\bar{t}) - x^u(\bar{t})) - [q(\bar{t}, x^{u+d}(\bar{t})) - q(\bar{t}, x^u(\bar{t}))] + r o_1(\|d\|_{\mathcal{L}^\infty})$$

(for some r independent of u and $o_1(\cdot)$ as in *Proposition 1.2*)

$$= [q_x(\bar{t}, x^u(\bar{t})) - q_x(\bar{t}, x^u(\bar{t}) + \xi)]\,(x^{u+d}(\bar{t}) - x^u(\bar{t})) + r o_1(\|d\|_{\mathcal{L}^\infty})$$

for some $\xi \in \mathcal{R}^n$ satisfying $\|\xi\| \leq \|x^{u+d} - x^u\|_{\mathcal{L}^\infty}$ by the Mean Value Theorem. But

$$\left\|x^{u+d} - x^u\right\|_{\mathcal{L}^\infty} \;\leq\; c_2\,\|d\|_{\mathcal{L}^2} \leq c_2\,\|d\|_{\mathcal{L}^\infty}$$

(c_2 as in *Proposition 1.1*). We deduce from the uniform continuity of q and q_x on bounded sets and the uniform boundedness of $\mathcal{U}$ that there exists $\theta : (0,\infty) \to (0,\infty)$ such that $\theta(s) \to 0$ as $s \downarrow 0$ and

$$\|q_x(t, x^v(t)) - q_x(t, x^v(t) + \eta)\| \leq \theta(\|e\|_{\mathcal{L}^\infty})$$

for all $t \in T$, $v \in \mathcal{U}$, $e \in \mathcal{L}_m^\infty[T]$ and $\eta \in \mathcal{R}^n$ such that $\|\eta\| \leq \|x^{v+e} - x^v\|_{\mathcal{L}^\infty}$. It follows that

$$\begin{aligned}\max_{t\in T} \Delta q^{u,d}(t) - \max_{t\in T} q(t, x^{u+d}(t)) \;&\leq\\ \theta(\|d\|_{\mathcal{L}^\infty})\,\left\|x^{u+d} - x^u\right\|_{\mathcal{L}^\infty} + r o_1(\|d\|_{\mathcal{L}^\infty}) \;&\leq\\ c_2\theta(\|d\|_{\mathcal{L}^\infty})\,\|d\|_{\mathcal{L}^\infty} + r o_1(\|d\|_{\mathcal{L}^\infty}) \;&=\; \tilde{o}(\|d\|_{\mathcal{L}^\infty})\end{aligned} \tag{1.7}$$

where $\tilde{o}(s) = c_2\theta(s)s + r o_1(s)$.

The inequality with Δq and q switched around is obtained by repeating the above arguments with $\bar{t}$ taken to be a minimizer of $q(\cdot, x^{u+d}(\cdot))$ in place of $\Delta q^{u,d}(\cdot)$. By (1.7)

$$\left|\max_{t\in R_{\varepsilon,u}} \Delta q^{u,d}(t) - \max_{t\in T} q(t, x^{u+d}(t))\right| \leq o_2^\varepsilon(\|d\|_{\mathcal{L}^\infty})$$

for all $u \in \mathcal{U}$ and $d \in \mathcal{L}_m^\infty[T]$, where

$$o_2^\varepsilon(s) = \begin{cases} \tilde{o}(s) & \text{for } s \in [0, \delta_\varepsilon] \\ m(s) & \text{for } s > \delta_\varepsilon. \end{cases}$$

Here $m(s)$ is a bound on the left side of (1.5) as (u, d) ranges over $\{(u, d) \in \mathcal{U} \times \mathcal{L}_m^\infty[T] : ||d||_{\mathcal{L}^\infty} \leq s\}$. $o_2^\varepsilon(\cdot)$ has the required properties. The proposition is proved. ■

Proposition 1.5 *Assume* **(H1)**. *Take a continuously differentiable function $h : \mathcal{R}^n \to \mathcal{R}$. Then there exists $o_3(\cdot) : (0, \infty) \to (0, \infty)$ such that $s^{-1} o_3(s) \to 0$ as $s \downarrow 0$ and*

$$\left| h(x^{u+d}(1)) - \left[h(x^u(1)) + h_x\left(x^u(1)\right) y^{u,d}(1) \right] \right| \leq o_3\left(||d||_{\mathcal{L}^\infty}\right)$$

for all $u \in \mathcal{U}$ and $d \in \mathcal{L}_m^\infty[T]$.

Proof.

The proof, based on *Proposition 1.3* and the Mean Value Theorem, involves a similar (but simpler) argument to that employed in the proof of *Proposition 1.4*. ■

Proposition 1.6 *Assume* **(H1)**. *Take continuously differentiable functions $h_i : \mathcal{R}^n \to \mathcal{R}$, $i = 1, 2, \ldots, k$. Then there exists $o_4(\cdot) : (0, \infty) \to (0, \infty)$ such that $s^{-1} o_4(s) \to 0$ as $s \downarrow 0$ and*

$$\left| \max_i \left| h_i(x^u(1)) + (h_i)_x(x^u(1)) y^{u,d}(1) \right| - \max_i \left| h_i(x^{u+d}(1)) \right| \right|$$
$$\leq o_4\left(||d||_{\mathcal{L}^\infty}\right)$$

for all $u \in \mathcal{U}$, $d \in \mathcal{L}_m^\infty[T]$.

Proof.

Choose $u \in \mathcal{U}$ and $d \in \mathcal{L}_m^\infty[T]$. Let $\max_i |h_i(x^{u+d}(1))|$ be achieved at $i = j$. By the Mean Value Theorem

$$h_j(x^{u+d}(1)) = h_j(x^u(1)) + (h_j)_x(x^u(1) + \xi)(x^{u+d}(1) - x^u(1))$$

for some $\xi \in \mathcal{R}^n$ such that $||\xi|| \leq ||x^{u+d} - x^u||_{\mathcal{L}^\infty}$. But

$$\left| (h_j)_x(x^u(1)) y^{u,d}(1) - (h_j)_x(x^u(1))(x^{u+d}(1) - x^u(1)) \right| \leq r o_1\left(||d||_{\mathcal{L}^\infty}\right),$$

where $o_1(\cdot)$ is as in *Proposition 1.3* and r is a bound on $|(h_i)_x(x^u(1))|$ as u ranges over $\mathcal{U}$ and i over $\{1, 2, \ldots, k\}$. It follows

$$\begin{aligned}
&\max_i \left|h_i(x^{u+d}(1))\right| - \max_i \left|h_i(x^u(1)) + (h_i)_x(x^u(1))y^{u,d}(1)\right| \\
&\quad \le \left|h_j(x^{u+d}(1))\right| - \left|h_j(x^u(1)) + (h_j)_x(x^u(1))y^{u,d}(1)\right| \\
&\quad \le \left|h_j(x^u(1)) + (h_j)_x(x^u(1)+\xi)(x^{u+d}(1) - x^u(1))\right| \\
&\quad - \left|h_j(x^u(1)) + (h_j)_x(x^u(1))(x^{u+d}(1) - x^u(1))\right| + ro_1\left(\|d\|_{\mathcal{L}^\infty}\right) \\
&\le |(h_j)_x(x^u(1)) - (h_j)_x(x^u(1)+\xi)| \left\|x^{u+d}(1) - x^u(1)\right\| + ro_1\left(\|d\|_{\mathcal{L}^\infty}\right),
\end{aligned}$$

in which $\xi \in \mathcal{R}^n$ satisfies

$$\|\xi\| \le \left\|x^{u+d}(1) - x^u(1)\right\| \le c_2 \|d\|_{\mathcal{L}^\infty}$$

(c_2 as in *Proposition 1.1*). Arguing as in the proof of *Proposition 1.4* we show that there exists $\psi : (0, \infty) \to (0, \infty)$ such that $\psi(s) \to 0$ as $s \downarrow 0$ and

$$|(h_i)_x(x^v(1)) - (h_i)_x(x^v(1)+\xi)| \le \psi\left(\|d\|_{\mathcal{L}^\infty}\right)$$

for all $v \in \mathcal{U}$, $i \in \{1, 2, \ldots, k\}$, $d \in \mathcal{L}^\infty_m[T]$ and $\xi \in \mathcal{R}^n$ such that $\|\xi\| \le \|x^{v+d}(1) - x^v(1)\|$. It follows

$$\begin{aligned}
\max_i \left|h_i(x^{u+d}(1))\right| - \max_i \left|h_i(x^u(1)) + (h_i)_x(x^u(1))y^{u,d}(1)\right| &\le \\
\psi\left(\|d\|_{\mathcal{L}^\infty}\right)\left\|x^{u+d}(1) - x^u(1)\right\| + ro_1\left(\|d\|_{\mathcal{L}^\infty}\right) &\le \\
c_2\psi\left(\|d\|_{\mathcal{L}^\infty}\right)\|d\|_{\mathcal{L}^\infty} + ro_1\left(\|d\|_{\mathcal{L}^\infty}\right) &\le \\
o_3\left(\|d\|_{\mathcal{L}^\infty}\right)&
\end{aligned}$$

where $o_3(s) = c_2\psi(s)s + ro_1(s)$.

Repeating the preceding arguments with j a maximizer for $\max_i |h_i(x^u(1)) + (h_i)_x(x^u(1))\, y^{u,d}(1))|$ we recover the preceding inequality with the two terms on the left reversed. The inequality is therefore valid with 'absolute value' on the left side. The proposition is proved. ■

2 Lagrangian, Hamiltonian and Reduced Gradients

In this section we consider functionals dependent on x and u, where x is the unique solution to nonlinear equations defined by u. The section is based, to large extent, on [124] and we refer in our presentation to functionals and system equations defined on abstract spaces. For the definitions see, for example, [62].

Consider the following problem. We have the functional $Q(x, u)$ defined on some Banach spaces $\mathcal{X}$, $\mathcal{U}$. We know that x and u are tied by the equations

$$P(x, u) = 0, \tag{2.1}$$

where P is some operator from $\mathcal{X} \times \mathcal{U}$ to $\mathcal{X}$. Furthermore, we assume that the first and second order partial derivatives of P and Q exist and are continuous, that equations (2.1) can be solved with respect to x and that this relation can be described by the function W, i.e. $x = W(u)$. Therefore, the functional $Q(x,u)$ can be represented as a functional depending only on u, namely $Q(x,u) = Q(W(u),u) = J(u)$.

Having that we want to find the gradient (the Frechet derivative) of J with respect to u and the expression for

$$J_{uu}\delta u \tag{2.2}$$

where δu is the perturbation of u (compare with d in *Proposition 1.3*). (2.2) can be used to define the Hessian of J.

Introduce the Lagrangian for the functional $Q(x,u)$ and constraints (2.1)

$$L(x,u,p) = Q(x,u) + \langle p, P(x,u)\rangle_{\mathcal{X}}\,, \tag{2.3}$$

where p is a linear functional from the dual space $\mathcal{X}^\star$. Here, we also have $\langle p,\cdot\rangle_{\mathcal{X}} : \mathcal{X} \to \mathcal{R}$.

First of all, we show that $J_u^\star(u)$, the gradient of J, can be found by solving the following equations

$$P(x,u) = L_p^\star(x,u,p) = 0 \tag{2.4}$$

$$Q_x^\star(x,u) + P_x^\star(x,u)p = L_x^\star(x,u,p) = 0 \tag{2.5}$$

$$J_u^\star(u) = Q_u^\star(x,u) + P_u^\star(x,u)p = L_u^\star(x,u,p). \tag{2.6}$$

Here, for example, $P_x^\star$ is the adjoint operator to P_x which is a partial derivative of P with respect to x.

Equations (2.4) are called the system equations, equations (2.5) are the adjoint equations and equations (2.6) are the gradient equations.

Calculating $J_u^\star(u)$ is relatively easy. From the adjoint equations, if $(P_x^\star)^{-1}$ —the inverse operator to $P_x^\star$—exists, we can calculate p and substitute it to the gradient equations:

$$J_u^\star(u) = Q_u^\star(x,u) - P_u^\star(x,u)(P_x^\star)^{-1}(x,u)Q_x^\star(x,u). \tag{2.7}$$

It is easy to check that we would get the same formula for the 'reduced' gradient, if instead of using equations (2.4)–(2.6) we applied the Implicit Function Theorem to the systems equations.

Establishing the formula for (2.2) is more elaborate. After linearizing equations (2.4)–(2.6) with respect to all variables, we arrive at the equations

$$L_{px}\delta x + L_{pu}\delta u = 0 \tag{2.8}$$

$$L_{xp}\delta p + L_{xx}\delta x + L_{xu}\delta u = 0 \tag{2.9}$$

$$L_{up}\delta p + L_{ux}\delta x + L_{uu}\delta u = J_{uu}\delta u. \tag{2.10}$$

Here, for example, $L_{ux} = L_{ux}(x,u,p)$.

Equations (2.8)–(2.10) can be used to obtain the formula for (2.2). First, δx is calculated from equations (2.8)

$$\delta x = -L_{px}^{-1} L_{pu} \delta u, \tag{2.11}$$

and equations (2.9) are transformed into the equations

$$L_{xp}\delta p - L_{xx} L_{px}^{-1} L_{pu} \delta u + L_{xu} \delta u = 0. \tag{2.12}$$

Next, δp is calculated from (2.12),

$$\delta p = L_{xp}^{-1} \left[L_{xx} L_{px}^{-1} L_{pu} - L_{xu} \right] \delta u, \tag{2.13}$$

and substituted into equations (2.10) gives

$$J_{uu}(u)\delta u = \left[L_{up} L_{xp}^{-1} L_{xx} L_{px}^{-1} L_{pu} - L_{up} L_{xp}^{-1} L_{xu} - L_{ux} L_{px}^{-1} L_{pu} + L_{uu} \right] \delta u.$$

Furthermore, if we take into account the definition of L, the formula for $J_{uu}(u)\delta u$ can be written in the following way

$$\begin{aligned} J_{uu}(u)\delta u \;=\; & \Big[P_u^\star (P_x^\star)^{-1} L_{xx} \, (P_x^\star)^{-1} P_u^\star - P_u^\star (P_x^\star)^{-1} L_{xu} - \\ & L_{ux} (P_x^\star)^{-1} P_u^\star + L_{uu} \Big] \delta u. \end{aligned} \tag{2.14}$$

This formula shows that the evaluation of the reduced Hessian is a fairly complex task and, with the exception of the last term in (2.14), it requires inverting some operator. The rest of this section will be dedicated to a particular realization of P for which formulas (2.6) and (2.10) can be derived relatively easily.

Consider P and Q defined by the equations:

$$\begin{aligned} P(x,u) = 0 \Leftrightarrow \dot{x}(t) &= f(t, x(t), u(t)), \text{ a.e. on } T, \\ & x(0) = x_0, \end{aligned} \tag{2.15}$$

$$Q(x,u) = \phi(x(1)), \tag{2.16}$$

and assume that $u \in \mathcal{L}_m^\infty[T]$, $x \in \mathcal{W}_1^\infty[T] (= \mathcal{X})$. where $\mathcal{W}_1^\infty[T]$ is the Sobolev space of differentiable functions with essentially bounded derivatives. It is well-known that in this particular case of P evaluations of $J_u^\star$ and $J_{uu}\delta u$ is relatively easy. Furthermore, derivatives considered are Frechet derivatives ([124]).

The system equations can be written in the form

$$P(x,u)(t) = \begin{bmatrix} \dot{x}(t) - f(t, x(t), u(t)) \\ x(0) - x_0 \end{bmatrix} = 0 \tag{2.17}$$

and $L(x,u,p)$ as

$$\begin{aligned} L(x,u,p) &= Q(x,u) + \langle p, P(x,u)\rangle_{\mathcal{X}} \\ &= \phi(x(1)) + \langle \hat{p}, x(0) - x_0\rangle_{\mathcal{R}^n} + \\ &\quad \int_0^1 \langle p(t), \dot{x}(t) - f(t,x(t),u(t))\rangle_{\mathcal{R}^n}\, dt. \end{aligned} \tag{2.18}$$

Consider the term

$$\int_0^1 \langle p(t), \dot{x}(t)\rangle_{\mathcal{R}^n}\, dt.$$

From the differentiation by part formula we write

$$\begin{aligned} \int_0^1 \langle p(t), \dot{x}(t)\rangle_{\mathcal{R}^n}\, dt &= \langle p(1), x(1)\rangle_{\mathcal{R}^n} - \langle p(0), x(0)\rangle_{\mathcal{R}^n} - \\ &\quad \int_0^1 \langle \dot{p}(t), x(t)\rangle_{\mathcal{R}^n}\, dt, \end{aligned}$$

thus,

$$\begin{aligned} L(x,u,p) &= \phi(x(1)) + \langle p(1), x(1)\rangle_{\mathcal{R}^n} + \\ &\quad \langle \hat{p} - p(0), x(0)\rangle_{\mathcal{R}^n} - \langle \hat{p}, x_0\rangle_{\mathcal{R}^n} - \\ &\quad \int_0^1 [\langle \dot{p}(t), x(t)\rangle_{\mathcal{R}^n} + \langle p(t), f(t,x(t),u(t))\rangle_{\mathcal{R}^n}]\, dt. \end{aligned}$$

If we introduce the Hamiltonian for P

$$H(t,x(t),u(t),p(t)) = \langle p(t), f(t,x(t),u(t))\rangle_{\mathcal{R}^n}, \tag{2.19}$$

equations (2.4)–(2.6) can be written as follows

$$\begin{aligned} L_p^\star(x,u,p) = 0 \quad &\Leftrightarrow \quad \dot{x}(t) = H_p(t,x(t),u(t),p(t))^T, && (2.20) \\ & \qquad x(0) = x_0 && (2.21) \\ L_x^\star(x,u,p) = 0 \quad &\Leftrightarrow \quad \dot{p}(t) = -H_x(t,x(t),u(t),p(t))^T, && (2.22) \\ & \qquad p(1) = -\phi_x(x(1))^T, \ \hat{p} = p(0) && (2.23) \\ L_u^\star(x,u,p) = J_u^\star \quad &\Leftrightarrow \quad J_u^T(t) = -H_u(t,x(t),u(t),p(t))^T. && (2.24) \end{aligned}$$

Equations (2.8)–(2.10) for P, Q defined by (2.15)–(2.16) can be provided in a similar way:

$$\begin{aligned} \delta\dot{x}(t) &= H_{px}(t)\delta x(t) + H_{pu}(t)\delta u(t), \quad \delta x(0) = 0 && (2.25) \\ \delta\dot{p}(t) &= -H_{xp}(t)\delta p(t) - H_{xx}(t)\delta x(t) - H_{xu}(t)\delta u(t) && (2.26) \\ \delta p(1) &= -\phi_{xx}(x(1))\delta x(1) && (2.27) \\ (J_{uu}\delta u)(t) &= -H_{up}(t)\delta p(t) - H_{ux}(t)\delta x(t) - H_{uu}(t)\delta u(t). && (2.28) \end{aligned}$$

Here, for example, $H_{xu}(t) = H_{xu}(t, x(t), u(t), p(t))$.

Equation (2.28) contains the term directly related to δu and it has two terms defined by δp, δx that are also related, but indirectly, to δu.

In order to eliminate δp from equation (2.28) we introduce new variables which satisfy the matrix equations:

$$\begin{aligned} r(\tau) &= H_{up}(\tau) && (2.29)\\ \dot{r}(t) &= H_{xp}(t)^T r(t), \quad t \in [\tau, 1]. && (2.30) \end{aligned}$$

From equations (2.26)–(2.27) and (2.29)–(2.30) we have

$$\begin{aligned} \int_\tau^1 \left[r(t)^T \delta\dot{p}(t) + \dot{r}(t)^T \delta p(t)\right] dt &= \int_\tau^1 \left[-r(t)^T H_{xp}(t)\delta p(t) - \right.\\ &\quad r(t)^T H_{xx}(t)\delta x(t) - \\ &\quad r(t)^T H_{xu}(t)\delta u(t) + \\ &\quad \left.\delta p(t)^T H_{xp}^T(t) r(t)\right] dt \\ &= r(1)^T \delta p(1) - r(\tau)^T \delta p(\tau) \\ &= -r(1)^T \phi_{xx}(1)\delta x(1) - H_{up}(\tau)\delta p(\tau), \end{aligned} \qquad (2.31)$$

so, eventually, we have the relation

$$\begin{aligned} H_{up}(\tau)\delta p(\tau) &= \int_\tau^1 \left[r(t)^T H_{xx}(t)\delta x(t) + r(t)^T H_{xu}(t)\delta u(t)\right] dt \\ &\quad -r(1)^T \phi_{xx}(1)\delta x(1) \end{aligned} \qquad (2.32)$$

which implies

$$\begin{aligned} (J_{uu}\delta u)(\tau) &= -H_{ux}(\tau)\delta x(\tau) - H_{uu}(\tau)\delta u(\tau) - \\ &\quad \int_\tau^1 \left[r(t)^T H_{xx}(t)\delta x(t) + r(t)^T H_{xu}(t)\delta u(t)\right] dt + \\ &\quad r(1)^T \phi_{xx}(1)\delta x(1). \end{aligned} \qquad (2.33)$$

In order to get a final formula for the reduced Hessian we introduce a new set of matrix equations

$$\begin{aligned} s(1) &= -\phi_{xx}(1)^T r(1) && (2.34)\\ \dot{s}(t) &= -H_{px}(t)^T s(t) - H_{xx}(t)^T r(t), \quad t \in (\tau, 1) && (2.35)\\ \dot{s}(t) &= -H_{px}(t)^T s(t) - \delta_D(t-\tau) H_{ux}(t)^T, \quad t \in [0, \tau], && (2.36) \end{aligned}$$

in which $\delta_D(\cdot)$ is a delta Dirac function.

From (2.25), (2.34)–(2.36) we write

$$\begin{aligned}\int_0^1 \left[\delta\dot{x}(t)^T s(t) + \delta x(t)^T \dot{s}(t)\right] dt &= \int_0^1 [\delta x(t)^T H_{px}(t)^T s(t) + \\ &\quad \delta u(t)^T H_{pu}(t)^T s(t) - \\ &\quad \delta x(t)^T H_{px}(t)^T s(t)]\, dt \\ &\quad - \int_\tau^1 \left[\delta x(t)^T H_{xx}(t)^T r(t)\right] dt - \\ &\quad H_{ux}(\tau)\delta x(\tau) \\ &= \delta x(1)^T s(1) \\ &= -\delta x(1)^T \phi_{xx}(1) r(1).\end{aligned} \tag{2.37}$$

(Notice that $\delta x(0) = 0$.)

(2.37) can be stated as

$$\int_\tau^1 \left[\delta x(t)^T H_{xx}(t)^T r(t)\right] dt + H_{ux}(\tau)\delta x(\tau) =$$
$$\int_0^1 \left[\delta u(t)^T H_{pu}(t)^T s(t)\right] dt + \delta x(1)^T \phi_{xx}(1) r(1).$$

Eventually, we come to the relation

$$\begin{aligned}(J_{uu}\delta u)(\tau) &= -\int_0^1 s(t)^T H_{pu}(t)\delta u(t) dt - \\ &\quad \int_\tau^1 r(t)^T H_{xu}(t)\delta u(t) dt - \\ &\quad H_{uu}(\tau)\delta u(\tau)\end{aligned} \tag{2.38}$$

and from that we can derive the reduced Hessian.

If we assume that control functions u are substituted by their piecewise-constant approximations defined on N subintervals by Nm values then, from formula (2.38), we can deduce that the reduced Hessian requires solving of

$$N(m+n)n/2 \tag{2.39}$$

linear differential equations (because the matrices involved are symmetric).

The above formulas for the reduced Hessian are rarely used since their evaluation is very expensive. Fortunately the reduced Hessian can be efficiently approximated by a quasi–Newton formula ([43]) as discussed in Chapter 5.

In Chapter 6 we present an approach to optimal control problems described by differential–algebraic equations. Its significant feature is that system equations (1.1) are discretized by an integration procedure with a variable stepsize. In this case we need a discrete–time model for (1.1).

Consider the operators $P(x,u)$ and $Q(x,u)$ defined as follows

$$\begin{aligned} P(x,u)=0 \Leftrightarrow x(k+1) &= f(k,x(k),u(k)),\ k=0,\ldots,N-1, \\ &\quad x(0)=x_0 \qquad (2.40) \\ Q(x,u) &= \phi(x(N)). \qquad (2.41) \end{aligned}$$

Here, $x(k)\in\mathcal{R}^n$, $u(k)\in\mathcal{R}^m$ and $\mathcal{X}=\mathcal{R}^{n(N+1)}$.

The Lagrangian for (2.40)–(2.41) is

$$\begin{aligned} L(x,u,p) &= Q(x,u)+\langle p,P(x,u)\rangle_{\mathcal{X}} \\ &= \phi(x(N))+\langle \hat{p},x(0)-x_0\rangle_{\mathcal{R}^n}+ \\ &\quad \sum_{k=0}^{N-1}\langle p(k+1),x(k+1)-f(k,x(k),u(k))\rangle_{\mathcal{R}^n}. \end{aligned}$$

If we introduce the notation

$$\begin{aligned} H(k,x(k),u(k),p(k+1)) &= \langle p(k+1),f(k,x(k),u(k))\rangle_{\mathcal{R}^n} \\ L(x,u,p) &= \phi(x(N))+\langle p(0),x(0)-x_0\rangle_{\mathcal{R}^n}- \\ &\quad \sum_{k=0}^{N-1}[H(k,x(k),u(k),p(k+1))- \\ &\quad \langle p(k+1),x(k+1)\rangle_{\mathcal{R}^n}], \end{aligned}$$

equations (2.4)–(2.6), for this particular case of P and Q, can be written as follows

$$\begin{aligned} L_p^\star(x,u,p)=0 &\Leftrightarrow x(k+1)=H_p(k,x(k),u(k),p(k+1))^T, \\ &\quad x(0)=x_0 \\ L_x^\star(x,u,p)=0 &\Leftrightarrow p(k)=H_x(k,x(k),u(k),p(k+1))^T, \qquad (2.42) \\ &\quad p(N)=-\phi_x(x(N))^T \qquad (2.43) \\ L_u^\star(x,u,p)=J_u^\star &\Leftrightarrow J_u^T(k)=-H_u(k,x(k),u(k),p(k+1))^T. \qquad (2.44) \end{aligned}$$

Similarly, equations (2.8)–(2.10) for the discrete–time problem are

$$\begin{aligned} \delta x(k+1) &= H_{px}(k)\delta x(k)+H_{pu}(k)\delta u(k),\quad \delta x(0)=0 \qquad (2.45) \\ \delta p(k) &= H_{xp}(k)\delta p(k)+H_{xx}(k)\delta x(k)+H_{xu}(k)\delta u(k) \qquad (2.46) \\ \delta p(N) &= -\phi_{xx}(x(N))\delta x(N) \qquad (2.47) \\ (J_{uu}\delta u)(k) &= -H_{up}(k)\delta p(k)-H_{ux}(k)\delta x(k)-H_{uu}(k)\delta u(k). \\ &\qquad (2.48) \end{aligned}$$

Here, $H_{xx}(k)$, $H_{xu}(k)$, etc, are evaluated at $(k,x(k),u(k),p(k+1))$.

If we introduce notation $\delta z(k) := (\{\delta x(k)\}_0^N, \{\delta u(k)\}_0^{N-1}, \{\delta p(k+1)\}_{-1}^{N-1})$, then linear equations (2.45)–(2.48) are sparse with respect to δz. This feature of the equations could be exploited while evaluating $(J_{uu}\delta u)(k)$. However, what we really need in numerical algorithms for optimal control problems (with possibly only simple constraints: $u \in \mathcal{U}$) is $J_{uu}^{-1}d$ where d is a descent direction for $J(u)$. Fortunately linear equations associated with this evaluation are also sparse ([34],[40],[84], see also [126],[61]).

Another approach to optimal control problems described by (2.40)–(2.41) is to use, together with the gradient formula (2.42)–(2.44), a general nonlinear programming algorithm suitable for large–scale problems. One such procedure, for problems with simple constraints, is described in [102].

3

First Order Method

Algorithms for optimal control problems can be analysed in the same way as algorithms for nonlinear programming problems. The compactness of the space of relaxed controls is only needed to guarantee boundedness of penalty parameters.

1 Introduction

This chapter concerns a first order, feasible directions algorithm for the solution of the following optimal control problem with pathwise state inequality constraints, labelled **(P)**:

$$\begin{aligned} &\min_u \phi(x(1)) \\ \text{s. t. } &\dot{x}(t) = f(t, x(t), u(t)), \text{ a.e. on } T, \; x(0) = x_0, \\ &u(t) \in \Omega \text{ a.e. on } T, \\ &h_i^1(x(1)) = 0 \; \forall i \in E, \\ &h_j^2(x(1)) \le 0 \; \forall j \in I \end{aligned} \tag{1.1}$$

and

$$q(t, x(t)) \le 0 \; \forall t \in T,$$

expressed in terms of the data: finite sets of index values E, I, functions $f : [0,1] \times \mathcal{R}^n \times \mathcal{R}^m \to \mathcal{R}^n$, $\phi : \mathcal{R}^n \to \mathcal{R}$, $h_i^1 : \mathcal{R}^n \to \mathcal{R}$ for $i \in E$, $h_j^2 : \mathcal{R}^n \to \mathcal{R}$ for $j \in I$ and $q : [0,1] \times \mathcal{R}^n \to \mathcal{R}$, a vector $x_0 \in \mathcal{R}^n$, and a set $\Omega \subset \mathcal{R}^m$ of the form

$$\Omega = \left\{ u \in \mathcal{R}^m : \; b_-^i \le u_i \le b_+^i \text{ for } i = 1, 2, \ldots, m \right\}, \tag{1.2}$$

in which b_-^i, b_+^i, $i = 1, 2, \ldots, m$ are constants. As in the previous chapter, throughout T denotes $[0, 1]$.

Everything that follows can be adapted to allow for multiple pathwise inequality constraints, for presence of an integral cost term and also for a general time horizon $[t_1, t_2]$; we limit ourselves to the above special case for notational simplicity.

The control problem **(P)** can be expressed as an optimization problem over the set of control functions

$$\mathcal{U} = \left\{ v \in \mathcal{L}_m^1[T] : \quad v(t) \in \Omega \text{ a. e. on } T \right\}$$

with the aid of the functions $\tilde{F}_0 : \mathcal{L}_m^2[T] \to \mathcal{R}$, $\tilde{h}_i^1 : \mathcal{L}_m^2[T] \to \mathcal{R}$ for $i \in E$, $\tilde{h}_j^2 : \mathcal{L}_m^2[T] \to \mathcal{R}$ for $j \in I$ and $\tilde{q} : \mathcal{L}_m^2[T] \to \mathcal{C}[T]$:

$$\begin{aligned} \tilde{F}_0(u) &= \phi(x^u(1)) \\ \tilde{h}_i^1(u) &= h_i^1(x^u(1)) \; \forall i \in E \\ \tilde{h}_j^2(u) &= h_j^2(x^u(1)) \; \forall j \in I \\ \tilde{q}(u)(t) &= q(t, x^u(t)) \; \forall t \in T. \end{aligned}$$

The reformulated problem is

$$\min_{u \in \mathcal{U}} \tilde{F}_0(u) \tag{1.3}$$

subject to

$$\begin{aligned} \tilde{h}_i^1(u) &= 0 \; \forall i \in E && (1.4) \\ \tilde{h}_j^2(u) &\leq 0 \; \forall j \in I && (1.5) \\ \tilde{q}(u)(t) &\leq 0 \; \forall t \in T. && (1.6) \end{aligned}$$

The algorithm we propose in this chapter is a feasible directions algorithm which applies an exact penalty function to equality constraints. However, the convergence analysis which we present here can also be applied in the context of other numerical methods intended for problems with the Chebyshev functional:

$$\tilde{Q}(u) = \max_{t \in T} \tilde{q}(u)(t). \tag{1.7}$$

The notable feature of this functional is the fact that the Dini derivative

$$\lim_{\varepsilon \to 0+} \frac{\tilde{Q}(u + \varepsilon h) - \tilde{Q}(u)}{\varepsilon} := d\tilde{Q}(u, h) \tag{1.8}$$

exists and can be expressed by the formula

$$d\tilde{Q}(u, h) = \max_{t \in R_{0,u}} q(t, x^u(t)) y^{u,h}(t), \tag{1.9}$$

where $y^{u,h}$ is the solution to the linearized equations defined by u and h (c.f. *Proposition 2.1.4*). Here,

$$R_{0,u} = \left\{ t \in T : \; \tilde{q}(u)(t) = \max_{\tilde{t} \in T} \tilde{q}(u)(\tilde{t}) \right\}.$$

Formula (1.9) holds under relatively general assumptions ([49], [42]). Although we never refer directly to formula (1.9) in our analysis, this feature of the Chebyshev functional is essential in establishing global convergence properties of several methods presented in this book.

The first order algorithm which we propose in this chapter has the following features:
a) the algorithm aims to solve a related problem $(\mathbf{P_c})$ in which the equality constraints are replaced by an 'exact penalty term' in the cost:

$$\min_{u \in \mathcal{U}} \tilde{F}_c(u)$$

subject to the constraints

$$\begin{aligned} \tilde{h}_j(u) &\leq 0 \ \forall j \in I \\ \tilde{q}(u)(t) &\leq 0 \ \forall t \in T \end{aligned}$$

in which

$$\tilde{F}_c(u) = \tilde{F}_0(u)/c + \max_{i \in E} \left| \tilde{h}_i^1(u) \right| .$$

b) The algorithm generates a sequence of controls whose corresponding state trajectories satisfy the pathwise and endpoint inequality constraints. Search directions are generated by solving a convex control subproblem. The new control is found by conducting a line search along the direction obtained from a direction finding subproblem.

Algorithms involving function space iterations for the solution of optimal control problems with pathwise state inequality constraints, with accompanying convergence analysis, have been proposed by Warga, Mayne and Polak and Polak, Yang and Mayne in [123], [75] and [86] respectively.

Both the Warga and Mayne–Polak algorithms involve proximity–type subalgorithms to generate search directions, and their effective implementation is hampered by the poor performance of proximity–type algorithms applied to the convex sets in infinite dimensional spaces which arise in this context. The algorithm of Warga generates a sequence of relaxed controls, accumulation points of which are shown to satisfy a strong version of the relaxed maximum principle. None of these algorithms involves an ε–active strategy for state constraints, a feature of our algorithm which greatly enhances its efficiency.

In the later Polak–Yang–Mayne algorithm, barrier functions are used to eliminate pathwise state constraints from the direction finding problems. Computation experience of this algorithm is limited, though preliminary findings are promising ([86]). The algorithm applies only to problems with no equality constraints. Furthermore, the associated unconstrained optimal control problems must be solved by the Newton method since they are ill–conditioned ([93]).

Machielsen ([72]), Alt and Malanowski ([2]) investigate 'function space' second order methods for solving optimal control problems with state constraints; a local convergence analysis and numerical examples are to be found in [2] and [72] respectively. The fact that the direction finding subproblems of [72] and [2] are, in general, nonconvex optimal control problems creates difficulties both regarding efficient implementation and global convergence analysis. A second order method along the lines of the approach given in this chapter is presented in Chapter 5.

Chapter 4 provides a full discussion of implementational aspects of the algorithm and also numerical examples. The examples include an optimal control problem arising in flight mechanics, concerning optimal control strategies in the presence of windshear, extensively studied by Bulirsh *et al* ([27],[28]). The fact that our feasible directions algorithm provides a solution to the 'windshear' problem without recourse to prior information about junction times or control structure (which are required in the method employed in [28]), is evidence of the effectiveness of our algorithm (and nonlinear programming methods in general) as a computational tool.

What special characteristics of the algorithm provided in this chapter promote efficient implementation? One is that search directions generated by the algorithm drive state trajectories into the interior of the state constraint region. This means that satisfaction of the state constraint over the entire time interval T can still be guaranteed, even if we impose the state constraint at only relatively few points in T (c.f. §4.1). Consequently a coarser discretization can be applied to the state constraint than that associated with the parametrization of control functions. This is significant since it is precisely the 'dimensionality' of the pathwise constraint which makes it difficult to compute optimal controls for the problem (**P**). Techniques for the approximation of sets on which the state constraint is required to be satisfied were anticipated in an algorithm proposed by Fedorenko ([42]). These techniques were also used in the algorithm proposed in [104], [105].

While techniques for deriving conditions on accumulation points generated in both feasible directions and also exact penalty methods in finite dimensional nonlinear programming are now available and well understood, developing a convergence analysis for the optimal control problem (**P**) poses additional difficulties, notably those associated with an inequality constraint function having infinite dimensional range and with the fact that the set $\mathcal{U}$ is not compact.

Novel features of the convergence analysis are as follows. We propose the convergence analysis based on 'nonpositive descent functions' which along the sequences generated by our algorithm is convergent. Its limit point, equal to zero, is the statement of necessary optimality conditions. A customary result in the literature would be that 'relaxed' accumulation points of sequences generated by our algorithm satisfy necessary conditions of optimality in the form of a 'relaxed' version of the maximum principle. Our convergence result is stronger in the sense that it is valid for the whole sequence generated by our algorithm. The compactness of the space of relaxed controls is only needed to guarantee boundedness of penalty parameters. The algorithm allows for a computation–saving ε–active strategy in dealing with the 'infinite dimensional' inequality constraint. A single, simply stated constraint qualification (hypothesis (**CQ**) in §3.4) is invoked both to ensure finite increase of the penalty parameter and derive properties of accumulation points, in place of a pair of constraint qualification type hypotheses featuring, for example, in [73], [75]. The relationship between standard necessary conditions of optimality and the 'nonpositive descent function' type conditions is also clarified.

Analytical techniques developed in this chapter are of benefit also in studying the convergence properties of related algorithms for solving optimal control problems, involving Chebyshev type functional constraints where, owing to the use of a variable stepsize in integration or high order integration procedures, it is either not possible or inconvenient to base the analysis on an *a priori* discretization of the dynamic equations. These issues are further discussed in Chapters 5–6. The algorithm (and its convergence analysis) presented in this chapter can be easily adapted to control problems with state constraints whose control functions are defined by piecewise constant (c.f. Chapters 5–6), or piecewise polynomial functions.

2 Representation of Functional Directional Derivatives

At each iteration of the feasible directions algorithm search directions are generated by solving a simplified version of the exact penalty function problem in which the dynamics, cost functional and constraint functionals are replaced by their first order approximations around the current control function u.

For $d \in \mathcal{U} - u$ we need to consider the first order approximation $x^u + y^{u,d}$ to x^{u+d} in which the perturbation $y^{u,d}$ to the nominal state trajectory x^u is the unique solution to the linearized equations

$$\begin{aligned} \dot{y}(t) &= f_x(t, x^u(t), u(t))y(t) + f_u(t, x^u(t), u(t))d(t), \text{ a.e. on } T, \\ y(0) &= 0. \end{aligned} \tag{2.1}$$

First order approximations to the functionals $\tilde{F}_0(u+d) - \tilde{F}_0(u)$, $\tilde{h}_i^1(u+d) - \tilde{h}_i^1(u)$, $i \in E$, $\tilde{h}_j^2(u+d) - \tilde{h}_j^2(u)$, $j \in I$, and $\tilde{q}(u+d)(t) - \tilde{q}(u)(t)$, $t \in T$, can now be defined via $y^{u,d}$ as follows:

$$\begin{aligned} \left\langle \nabla \tilde{F}_0(u), d \right\rangle &:= \phi_x(x^u(1))y^{u,d}(1), \\ \left\langle \nabla \tilde{h}_i^1(u), d \right\rangle &:= (h_i^1)_x(x^u(1))y^{u,d}(1) \text{ for } i \in E, \end{aligned}$$

$$\left\langle \nabla \tilde{h}_j^2(u), d \right\rangle := (h_j^2)_x(x^u(1))y^{u,d}(1) \text{ for } j \in I,$$

and

$$\langle \nabla \tilde{q}(u)(t), d \rangle := q_x(t, x^u(t))y^{u,d}(t) \text{ for } t \in T.$$

The notation $\langle \nabla \tilde{F}_0(u), d \rangle$ is intended to convey the suggestion that $\langle \nabla \tilde{F}_0 (u), \cdot \rangle$ is a directional derivative associated with some kind of 'derivative' $\nabla \tilde{F}_0$ of the functional $\tilde{F}_0$ at u. It is however unnecessary to pursue this interpretation (this would require us to specify function spaces and the precise notation of the derivative $\nabla \tilde{F}_0$). As far as describing numerical methods and analysing their convergence properties are concerned, the simplest course is to take the above formulas for $\langle \nabla \tilde{F}_0(u), d \rangle$, etc, as *definitions* of constructs featuring in the algorithm whose properties can be analysed directly with the help of the results on the relationship between nonlinear control systems and their linear approximations as discussed in Chapter 2.

3 Relaxed Controls

The convergence analysis to follow involves relaxed controls. In this section we briefly review their properties and introduce some notation.

We recall that a *Radon probability measure* ς on the Borel sets of Ω is a regular positive measure such that $\varsigma(\Omega) = 1$. The set of all Radon probability measures is denoted by $rpm(\Omega)$. A *relaxed control*, μ, is a measurable function $\mu : T \to rpm(\Omega)$ where 'measurability' is as defined in [121]. The set of relaxed controls is denoted by $\bar{\mathcal{U}}$.

Let $\mathcal{L}^1(T, \mathcal{C}(\Omega))$ denote the space of absolutely integrable functions from T to $\mathcal{C}(\Omega)$. Then the topology imposed on $\bar{\mathcal{U}}$ is the weakest topology such that the mapping

$$\mu \to \int_T \int_\Omega \psi(t, u)\mu(t)(du)dt$$

is continuous for all $\psi \in \mathcal{L}^1(T, \mathcal{C}(\Omega))$.

We recall ([121], p. 287) that $\bar{\mathcal{U}}$ is a compact and convex subset of a normed vector space (namely the dual space of $\mathcal{L}^1(T, \mathcal{C}(\Omega))$ with a 'weak' norm whose topology restricted to $\bar{\mathcal{U}}$ coincides with the weak star topology).

Relaxed controls give rise to the relaxed dynamics:

$$\dot{x}(t) = f_r(t, x(t), \mu(t)) := \int_\Omega f(t, x(t), u)\mu(t)(du), \ x(0) = x_0$$

whose solution we denote by x_r^μ. Extensions to relaxed controls of functions in the problem (**P**) are denoted as follows: $\hat{F}_0(\mu) = \phi(x_r^\mu(1))$, $\hat{h}_i^1(\mu) =$

$h_i^1(x_r^\mu(1))$, $\hat{h}_j^2(\mu) = h_j^2(x_r^\mu(1))$, $\hat{q}(\mu)(t) = q(t, x_r^\mu(t))$. We can define then the relaxed problem ($\mathbf{P^r}$):

$$\min_{\mu \in \bar{\mathcal{U}}} \hat{F}_0(\mu) \tag{3.1}$$

subject to

$$\begin{aligned} \hat{h}_i^1(\mu) &= 0 \;\forall i \in E && (3.2)\\ \hat{h}_j^2(\mu) &\leq 0 \;\forall j \in I && (3.3)\\ \hat{q}(\mu)(t) &\leq 0 \;\forall t \in T. && (3.4) \end{aligned}$$

With each ordinary control $u \in \mathcal{U}$ we associate a relaxed control $\mu \in \bar{\mathcal{U}}$ defined by the property $\mu(t)(S) = \delta_{u(t)}(S)$ for all Borel sets $S \subset \Omega$, where $\delta_u(S) = 1$ if $u \in S$ and $\delta_u(S) = 0$ otherwise. We write this control $\mathbf{u}$. Naturally, u and $\mathbf{u}$ give rise to the same state trajectory.

A useful concept, introduced in [5], for studying approximations to relaxed state trajectories, is the set of *search directions*:

$$\mathcal{D} := \{d(\cdot,\cdot) : T \times \Omega \to \mathcal{R}^m : \; d(\cdot, u) \text{ is measurable, } d(t, \cdot) \text{ is continuous and } u + d(t,u) \in \Omega \;\forall u \in \Omega, \text{ a.e. on } T\}$$

Take $\mu \in \bar{\mathcal{U}}$ and $d \in \mathcal{D}$. $y_r^{\mu,d}$ is the solution to the relaxed version of (2.1):

$$\begin{aligned} \dot{y}(t) &= (f_x)_r(t, x_r^\mu(t), \mu(t))y(t) + \int_\Omega f_u(t, x_r^\mu(t), u)d(t,u)\mu(t)(du),\\ y(0) &= 0. \end{aligned}$$

Note that if $\mu(t) = \mathbf{u}(t)$ and $d(t,u) = v(t) - u$ for some $v \in \mathcal{U}$ then $y_r^{\mu,d} = y^{u,v-u}$.

For $\mu \in \bar{\mathcal{U}}$ and $d \in \mathcal{D}$ we denote

$$\begin{aligned} \left\langle \nabla \hat{F}_0(\mu), d \right\rangle_r &:= \phi_x(x_r^\mu(1))^T y_r^{\mu,d}(1)\\ \left\langle \nabla \hat{h}_i^1(\mu), d \right\rangle_r &:= (h_i^1)_x(x_r^\mu(1)) y_r^{\mu,d}(1) \text{ for } i \in E\\ \left\langle \nabla \hat{h}_j^2(\mu), d \right\rangle_r &:= (h_j^2)_x(x_r^\mu(1)) y_r^{\mu,d}(1) \text{ for } j \in I \end{aligned}$$

and

$$\langle \nabla \hat{q}(\mu)(t), d \rangle_r := q_x(t, x_r^\mu(t)) y_r^{\mu,d}(t) \text{ for } t \in T.$$

Under the hypotheses **(H1)** and **(H2)** (to be imposed in §3.4) we deduce from standard properties of relaxed controls ([121]): for fixed $d \in \mathcal{D}$, the following mappings are continuous

$$\mu \to x_r^\mu, \;\; \mu \to \left\langle \nabla \hat{F}_0(\mu), d \right\rangle_r, \;\; \mu \to \left\langle \nabla \hat{h}_i^1(\mu), d \right\rangle_r, \; i \in E,$$
$$\mu \to \left\langle \nabla \hat{h}_j^2(\mu), d \right\rangle_r, \; j \in I, \;\; \mu \to \langle \nabla \hat{q}(\mu)(\cdot), d \rangle_r$$

where the domain in each case is $\bar{\mathcal{U}}$ and the range spaces $\mathcal{C}(T, \mathcal{R}^n)$, $\mathcal{R}^{|E|}$, $\mathcal{R}^{|I|}$ and $\mathcal{C}(T)$ respectively.

4 The Algorithm

We begin by describing the direction finding subproblem and some functions associated with it.

For fixed c and u the direction finding subproblem $\mathbf{P_c(u)}$ for problem $(\mathbf{P_c})$ is:

$$\min_{d \in \mathcal{U}-u, \beta \in \mathcal{R}} \beta + 1/(2c) \, ||d||^2_{\mathcal{L}^2}$$

subject to

$$\begin{aligned}
\left\langle \nabla \tilde{F}_0(u), d \right\rangle /c + \max_{i \in E} \left| \tilde{h}^1_i(u) + \left\langle \nabla \tilde{h}^1_i(u), d \right\rangle \right| - \max_{i \in E} \left| \tilde{h}^1_i(u) \right| &\leq \beta \\
\tilde{h}^2_j(u) + \left\langle \nabla \tilde{h}^2_j(u), d \right\rangle &\leq \beta, \\
&\quad \forall j \in I \\
\tilde{q}(u)(t) + \langle \nabla \tilde{q}(u)(t), d \rangle &\leq \beta, \\
&\quad \forall t \in R_{\varepsilon, u}.
\end{aligned}$$

Here $\varepsilon > 0$ is a parameter which governs the tightness of the approximation to the active region of the pathwise inequality constraint (c.f. (2.1.6)).

The subproblem can be reformulated as an optimization problem over the space $\mathcal{L}^2_m[T]$ whose objective function is strictly convex. It therefore has a unique solution $(\bar{d}, \bar{\beta})$. Since this solution depends on c and u, we may define the *descent function* $\sigma_c(u)$ and the *penalty test function* $t_c(u)$, which will be used to test optimality of a control u and to adjust c, respectively, as

$$\sigma_c(u) = \bar{\beta}$$

and

$$t_c(u) = \bar{\beta} + \max_{i \in E} \left| \tilde{h}^1_i(u) \right| /c$$

for given $c > 0$ and $u \in \mathcal{U}$.

A starting point is required which is feasible with respect to the inequality constraints. This point is computed by applying a few iterations of an algorithm which is a simple variant of the algorithm below (see, e.g. [107]). Other approaches are possible such as that described in [85].

The algorithm is as follows.

Algorithm 1 Fix parameters: $\varepsilon > 0$, γ, $\eta \in (0,1)$, $c^0 > 0$, $\kappa > 1$.

1. Choose the initial control $u_0 \in \mathcal{U}$ which satisfies $\tilde{h}_j^2(u_0) \leq 0 \; \forall j \in I$ and $\tilde{q}(u_0)(t) \leq 0 \; \forall t \in T$. Set $k = 0$, $c_{-1} = c^0$.

2. Let c_k be the smallest number chosen from $\{c_{k-1}, \kappa c_{k-1}, \kappa^2 c_{k-1}, \ldots\}$ such that the solution (d_k, β_k) to the direction finding subproblem $\mathbf{P}_{\mathbf{c_k}}(\mathbf{u_k})$ satisfies

$$t_{c_k}(u_k) \leq 0.$$

If $\sigma_{c_k}(u_k) = 0$ then STOP.

3. Let α_k be the largest number chosen from the set $\{1, \eta, \eta^2, \ldots,\}$ such that $u_{k+1} = u_k + \alpha_k d_k$ satisfies the relations

$$\begin{aligned} \tilde{F}_{c_k}(u_{k+1}) - \tilde{F}_{c_k}(u_k) &\leq \gamma \alpha_k \sigma_{c_k}(u_k) \\ \tilde{h}_j^2(u_{k+1}) &\leq 0 \; \forall j \in I \\ \tilde{q}(u_{k+1})(t) &\leq 0 \; \forall t \in T. \end{aligned}$$

Increase k by one. Go to Step 2.

The descent function $\sigma_{c_k}(u_k)$ is nonpositive valued at each iteration. Suppose that subsequences $\{u_k\}_{k \in K}$, $\{c_k\}_{k \in K}$ of the sequences of control functions and penalty parameters generated by the algorithm have limit points (in some sense) $\bar{u}$ and $\bar{c}$. We would then expect that $\sigma_{c_k}(u_k) \to \sigma_{\bar{c}}(\bar{u})$ (along the subsequence) and $\sigma_{\bar{c}}(\bar{u}) \geq 0$, a condition which asserts that the direction finding subproblem for $c_k = \bar{c}$ and $u_k = \bar{u}$, has the solution $(d = 0, \beta = 0)$ and which (together with the feasibility of $\bar{u}$) can be interpreted as a first order optimality condition satisfied by $\bar{u}$. If $\{c_k\}$ is nondecreasing and bounded (thus convergent) then $\sigma_{c_k}(u_k) \to_{k \to \infty} 0$ would be a stronger result because the existence of convergent subsequence of $\{u_k\}$ is not requested. Justifying these arguments (under precisely specified hypotheses) is the essence of the convergence analysis to follow.

If u is not a stationary point for the problem (**P**) then $d \neq 0$ and $\beta < 0$ which implies that

$$\tilde{q}(u)(t) + \langle \nabla \tilde{q}(u)(t), d \rangle < 0 \; \forall t \in R_{\varepsilon,u}. \tag{4.1}$$

We can show, under the continuity assumption imposed on $q_x(\cdot,\cdot)$, that $\tilde{q}(u)(t) + \langle \nabla \tilde{q}(u)(t), d \rangle$ is Lipschitz function with respect to t (§4.1), therefore there is a finite set of points $A \subset R_{\varepsilon,u}$ such that, if

$$\tilde{q}(u)(t) + \langle \nabla \tilde{q}(u)(t), d \rangle < 0 \; \forall t \in A$$

then (4.1) is also satisfied. This property is exploited in the implementable version of the algorithm discussed in Chapter 4.

Note that the only requirement in the directional minimization related to the state constraint is the feasibility of u_{k+1}. Other exact penalty function methods such as that in [75] do not combine the strict descent property (4.1) with the unrestrictive direction minimization procedure regarding the state constraint.

The role of the penalty parameter test function t_c is to ensure that the penalty parameter is large enough in the limit (but finite) to force satisfaction of the equality endpoint constraint. The algorithm ensures that, at the kth iteration,

$$t_{c_k}(u_k) = \sigma_{c_k}(u_k) + \max_{i \in E} \left| \tilde{h}_i^1(u_k) \right| / c_k \le 0.$$

Since under favorable circumstances $\sigma_{c_k}(u_k) \to 0$ and $\{c_k\}$ is bounded we conclude from this inequality that $\max_{i \in E} |\tilde{h}_i^1(\bar{u})| = 0$, i.e. the limiting control satisfies the endpoint equality constraint.

Before concluding this section we need to clarify two points about the algorithm. They are to guarantee that Step 2 and 3 of *Algorithm 1* can always be carried out. These gaps are filled by the following proposition in which we invoke additional hypotheses including a constraint qualification.

(H2): ϕ, h_i^1, $i \in E$, h_j^2, $j \in I$ are continuously differentiable functions. $q(t, \cdot)$ is differentiable for each t and q, q_x are continuous functions.

(CQ): For each $\mu \in \bar{\mathcal{U}}$ we have $\mathcal{F}(\mu) \ne \emptyset$ and, in the case $E \ne \emptyset$,

$$0 \in \text{interior}[\mathcal{E}(\mu)],$$

where

$$\mathcal{E}(\mu) := \left\{ \left\{ \left\langle \nabla \hat{h}_i^1(\mu), d \right\rangle_r \right\}_{i \in E} \in \mathcal{R}^{|E|} : d \in \mathcal{F}(\mu) \right\}$$

and

$$\mathcal{F}(\mu) := \left\{ d \in \mathcal{D} : \max_{j \in I} \left[\min \left[0, \hat{h}_j^2(\mu) \right] + \left\langle \nabla \hat{h}_j^2(\mu), d \right\rangle_r \right] < 0, \right.$$

$$\left. \max_{t \in T} [\min [0, \hat{q}(\mu)(t)] + \langle \nabla \hat{q}(\mu)(t), d \rangle_r] < 0 \right\}.$$

In the case $I = \emptyset$ (no terminal equality constraints), we interpret $\max_{j \in I} [\tilde{h}_j^2(\mu) + \cdots] = -\infty$.

(CQ) is a constraint qualification related to hypotheses earlier invoked by Mayne and Polak ([73]). It is a local controllability condition on values of the (linearized) equality constraint functions with respect to control functions which are strictly feasible w.r.t. the linearized inequality constraints. Note that $\mathcal{F}(\mu)$ can be stated, for the feasible directions algorithm, alternatively as:

$$\mathcal{F}(\mu) = \Big\{ d \in \mathcal{D} : \max_{j \in I} \left[\hat{h}_j^2(\mu) + \left\langle \nabla \hat{h}_j^2(\mu), d \right\rangle_r \right] < 0,$$

$$\max_{t \in T} \left[\hat{q}(\mu)(t) + \langle \nabla \hat{q}(\mu)(t), d \rangle_r \right] < 0 \Big\}.$$

To show that **(CQ)** is the generalization of well–known constraint qualifications (such as those stated in [12], [43] for nonlinear programming problems) consider the condition

$$0 \in \text{interior}\left\{ \{ \langle a_i^1, d \rangle_{\mathcal{R}^n} \}_{i \in E} : \langle a_j^2, d \rangle_{\mathcal{R}^n} < 0 \; \forall j \in I, \; d \in \mathcal{R}^n \right\}, \quad (4.2)$$

where $a_i^1, a_j^2 \in \mathcal{R}^n$, $i \in E$, $j \in I$. We can show that if a_i^1, $i \in E$, a_j^2, $j \in I$ are linearly independent then (4.2) holds. We will prove this for a simpler case when $I = \emptyset$ (the general case, $I \neq \emptyset$, requires a little more involved analysis).

Assume that a_i^1, $i \in E$ are linearly independent. We want to prove that (4.2) is true. If this is not the case there exist ψ_i, $i \in E$, not all equal to zero, such that

$$\sum_{i \in E} \psi_i \langle a_i^1, d \rangle_{\mathcal{R}^n} \geq 0 \quad \forall d \in \mathcal{R}^n \quad (4.3)$$

(this follows from a separation theorem stated, for instance, in [110]). Take $\hat{d} = -\sum_{i \in E} \psi_1 a_i^1$, then from (4.3) $||\hat{d}|| = 0$ which implies that $\psi_i = 0$, $i \in E$ since a_i^1, $i \in E$ are linearly independent. This is the contradiction, therefore (4.3) is true.

We can also prove the opposite. Indeed, if a_i^1, $i \in E$ are linearly dependent then there exists ψ_i, $\in E$, not all equal to zero, such that

$$\sum_{i \in E} \psi_i a_i^1 = 0.$$

This implies that

$$\sum_{i \in E} \psi_i \langle a_i^1, d \rangle_{\mathcal{R}^n} = \left\langle \sum_{i \in E} \psi_i a_i^1, d \right\rangle_{\mathcal{R}^n} \geq 0 \;\; \forall d \in \mathcal{R}^n.$$

However, the set $\{ \{ \langle a_i^1, d \rangle_{\mathcal{R}^n} \}_{i \in E} : \; d \in \mathcal{R}^n \}$ is convex and separated from the origin by the hyperplane defined by nonzero ψ_i, $i \in E$. This is the contradiction to (4.2) (see [110]).

The role of **(CQ)** is to ensure uniform boundedness of penalty parameter values and that a convergence analysis can be carried out in terms of 'normal' extremality conditions (conditions in which the cost multiplier is nonzero).

Proposition 4.1 *(i) Assume that hypotheses* **(H1)**, **(H2)** *and* **(CQ)** *are satisfied. Then for any $u \in \mathcal{U}$ satisfying the endpoint and pathwise inequality constraints of* **(P)** *there exists $\bar{c} > 0$ such that for all $c > \bar{c}$*

$$t_c(u) \leq 0.$$

(ii) Assume that hypotheses **(H1)** *and* **(H2)** *are satisfied. Then for any $u \in \mathcal{U}$, satisfying the endpoint and pathwise inequality constraints, and $c > 0$ such that $\sigma_c(u) < 0$ there exists $\bar{\alpha} > 0$ such that if $\alpha \in [0, \bar{\alpha})$ then*

$$\begin{aligned} \tilde{F}_c(\tilde{u}) - \tilde{F}_c(u) &\leq \gamma\alpha\sigma_c(u) \\ \tilde{h}_j^1(\tilde{u}) &\leq 0 \; \forall j \in I \\ \tilde{q}(\tilde{u})(t) &\leq 0 \; \forall t \in T, \end{aligned}$$

where $\tilde{u} = u + \alpha d$ and $(d, \beta = \sigma_c(u))$ is the solution to the direction finding subproblem corresponding to c and u.

A proof of this proposition is given in §3.6.

5 Convergence Properties of the Algorithm

In this section we show that the descent function $\sigma_c(u)$ is convergent to zero along the sequences $\{c_k\}$, $\{u_k\}$ generated by *Algorithm 1*. Furthermore, we show that $\{c_k\}$ is bounded and that any accumulation point of $\{u_k\}$ (regarded as a sequence in $\bar{\mathcal{U}}$) satisfies necessary optimality conditions.

Results are given in relation to the necessary conditions $(\mathbf{NC^r})$ in normal form for a control function $\bar{\mu}$, which is feasible for $(\mathbf{P^r})$, to be a minimizer.

$(\mathbf{NC^r})$: There exist nonnegative numbers α_j^2, $j \in I$, numbers α_i^1, $i \in E$, an absolutely continuous function $p_r : [0,1] \to \mathcal{R}^n$ and a nonnegative regular measure ν on the Borel subsets of $[0,1](= T)$ such that

$$-\dot{p}_r(t) = (f_x)_r(t, x_r^{\bar{\mu}}(t), \bar{\mu}(t))^T \left(p_r(t) + \int_{[0,t)} q_x(s, x_r^{\bar{\mu}}(s))^T \nu(ds) \right),$$

$$-\left(p_r(1)^T + \int_{[0,1]} q_x(s, x_r^{\bar{\mu}}(s))\nu(ds) \right) = \phi_x(x_r^{\bar{\mu}}(1)) +$$

$$\sum_{i \in E} \alpha_i^1 (h_i^1)_x(x_r^{\bar{\mu}}(1)) + \sum_{j \in I} \alpha_j^2 (h_j^2)_x(x_r^{\bar{\mu}}(1)),$$

$$\left(p_r(t)^T + \int_{[0,t)} q_x(s, x_r^{\bar{\mu}}(s))\nu(ds) \right) \cdot$$

$$\int_\Omega f_u(t, x_r^{\bar{\mu}}(t), u)d(t,u)\bar{\mu}(t)(du) \leq 0 \;\forall d \in \mathcal{D}, \text{ a.e. on } T,$$

$$\text{supp}\{\nu\} \subset \{t \in T : \; q(t, x_r^{\bar{\mu}}(t)) = 0\} \text{ and } \alpha_j^2 = 0 \text{ if } h_j^2(x_r^{\bar{\mu}}(1)) < 0. \tag{5.1}$$

Here supp$\{\nu\}$ denotes the support of the measure ν.

Conditions (**NC^r^**) are standard necessary optimality conditions for a relaxed minimizer (derivable for example from [121], Theorem VI.2.3) valid under hypotheses (**H1**), (**H2**) and (**CQ**) with the exception that (5.1) replaces the customary

$$\text{supp}\{\bar{\mu}(t)\} \subset \arg\max_{u\in\Omega} r(t)^T f(t, x_r^{\bar{\mu}}(t), u), \text{a.e. on } T, \tag{5.2}$$

where $r(t) = p_r(t) + \int_{[0,t)} q_x(s, x_r^{\bar{\mu}}(s))^T \nu(ds)$.

However, (5.2) implies that for all $d \in \mathcal{D}$ and $\varepsilon > 0$

$$\varepsilon^{-1} \int_\Omega r(t)^T \left(f(t, x_r^{\bar{\mu}}(t), u + \varepsilon d(t,u)) - f(t, x_r^{\bar{\mu}}(t), u)\right) \bar{\mu}(t)(du) \leq 0,$$

a.e. on T. Passing to the limit as $\varepsilon \downarrow 0$ with the help of the dominated convergence theorem gives (5.1). It follows that (**NC^r^**) are necessary conditions as claimed.

Theorem 5.1 *Assume that the data for* (**P**) *satisfies hypotheses* (**H1**), (**H2**) *and* (**CQ**). *Let* $\{u_k\}$ *be a sequence of control functions generated by* Algorithm 1 *and let* $\{c_k\}$ *be the corresponding sequence of penalty parameters. Then*

(i) $\{c_k\}$ *is a bounded sequence,*

(ii)

$$\lim_{k\to\infty} \sigma_{c_k}(u_k) = 0, \quad \lim_{k\to\infty} \max_{i\in E} \left|\tilde{h}_i^1(u_k)\right| = 0,$$

(iii) if $\bar{\mu}$ *is any accumulation point of* $\{\mathbf{u}_k\}$ *in* $\bar{\mathcal{U}}$ *(and such an accumulation point always exists) then* $\bar{\mu}$ *is feasible for the relaxed problem* (**P^r^**) *and satisfies necessary conditions* (**NC^r^**).

Note that if $\{u_k\}$ is a sequence in $\mathcal{U}$ such that $u_k \to \bar{u}$ for some $\bar{u} \in \mathcal{L}_m^2[T]$ with respect to the $\mathcal{L}^2$ norm then $\mathbf{u}_k \to \bar{\mathbf{u}}$ converges in $\bar{\mathcal{U}}$ ([125], see also [115]). Part *(iii)* of *Theorem 5.1* may therefore be substituted by the weaker assertion

(iii) if $\bar{u} \in \mathcal{U}$ is any $\mathcal{L}^2$ accumulation point of $\{u_k\}$ then $\bar{u}$ is feasible for (**P**) and satisfies (**NC^r^**).

Notice that the conditions ($\mathbf{NC^r}$), when applied at an ordinary control, reduce to simpler 'non–relaxed' necessary conditions of optimality (c.f. conditions (**NC**) in Chapter 4). We see then that the 'relaxed' analysis subsumes the $\mathcal{L}^2$ analysis, and improves on it by giving information about asymptotic behaviour of the algorithm even when $\mathcal{L}^2$ accumulation points do not exist.

Part *(iii)* of the theorem implies that if $u \in \mathcal{U}$ is feasible and $\sigma_c(u) = 0$ for some $c > 0$ then u satisfies necessary conditions ($\mathbf{NC^r}$) of optimality. Fix $\varepsilon_{\rm STOP} > 0$. Part *(ii)* of the theorem implies that the stopping condition

$$\sigma_{c_k}(u_k) \geq -\varepsilon_{\rm STOP}, \qquad \max_{i \in E} \left|\tilde{h}_i^1(u_k)\right| \leq \varepsilon_{\rm STOP},$$
$$\max_{j \in I} \tilde{h}_j^2(u_k) \leq 0, \qquad \max_{t \in T} \tilde{q}(u_k)(t) \leq 0.$$

is satisfied after a finite number of iterations. Termination of the algorithm still occurs after a finite number of iterations if the above stopping criterion is supplemented by

$$\left|\tilde{F}_0(u_{k+1}) - \tilde{F}_0(u_k)\right| \leq \varepsilon_{\rm STOP}, \quad ||u_{k+1} - u_k||_{\mathcal{L}^2} \leq ||d_k||_{\mathcal{L}^2} \leq \varepsilon_{\rm STOP}.$$

The first inequality follows from *(i)*, *(ii)* and the first inequality of *Proposition 4.1(ii)*. The second inequality is a consequence of the fact that the optimal value of the subproblem $\mathbf{P_c(u)}$ is nonpositive whence

$$||d_k||_{\mathcal{L}^2} \leq -2c_k\sigma_{c_k}(u_k) \text{ and } (ii) \implies \lim_{k\to\infty} ||d_k||_{\mathcal{L}^2} = 0.$$

Notice that *(i)* and *(ii)* correspond to the following general convergence result in nonlinear programming related to minimizing a bounded (from below), continuously differentiable function f: $\lim_{k\to\infty} \nabla f(x_k) = 0$. (The condition does not require the existence of accumulation points of $\{x_k\}$ and is relevant, for example, in situations when we seek to minimize $f(x) = e^x$.) We are not aware of convergence results, of this general nature, elsewhere in the constrained nonlinear programming literature. The condition *(ii)* was stated, for the first time, in [92].

The significance of our convergence results is that it is not necessary to introduce the space of relaxed controls to analyse convergence of optimal control algorithms.[1] In fact optimal control problems are 'well behaved' optimization problems—if we assume sufficiency optimality conditions at a point satisfying necessary optimality conditions we can also show that

[1]Notice that a standard assumption in nonlinear programming literature is that sequences generated by algorithms for constrained problems lie in compact sets—[12], [41]. In our convergence analysis the compactness of the space of relaxed controls is only needed to prove boundedness of the sequence of penalty parameters $\{c_k\}$.

this point will be a point of attraction for the sequence $\{u_k\}$ (see analysis in [2] and Chapter 5). This observation questions the practical justification for optimal control algorithms defined in the space of relaxed controls as described in [122] and [123].[2] However, we still have to bear in mind that when we solve an optimal control problem we aim at finding $\mathcal{L}_m^\infty[T]$ functions, a considerable task which we simplify by performing only a finite number of iterations of an optimal control procedure. This is often reflected in jittery behaviour of controls which only approximate to solutions.

6 Proof of the Convergence Theorem, etc.

We precede the proof of *Proposition 4.1* with a lemma which describes important implications of the constraint qualification (**CQ**).

Lemma 6.1 *Assume* (**H1**), (**H2**) *and* (**CQ**). *For any relaxed control* $\mu \in \bar{\mathcal{U}}$ *there exist a neighbourhood* $\mathcal{O}(\mu)$ *of* μ, *in the relaxed topology,* $K_1 > 0$ *and* $K_2 > 0$ *with the following properties: given any* $u \in \mathcal{U}$ *such that* $\mathbf{u} \in \mathcal{O}(\mu)$ *there exists* $v \in \mathcal{U}$ *such that*

$$\max_{i \in E}\left|\tilde{h}_i^1(u) + \left\langle \nabla \tilde{h}_i^1(u), v - u\right\rangle\right| - \max_{i \in E}\left|\tilde{h}_i^1(u)\right| \leq -K_1 \max_{i \in E}\left|\tilde{h}_i^1(u)\right| \tag{6.1}$$

$$\max_{j \in I}\left[\min\left[0, \tilde{h}_j^2(u)\right] + \left\langle \nabla \tilde{h}_j^2(u), v - u\right\rangle\right] \leq -K_1 \max_{i \in E}\left|\tilde{h}_i^1(u)\right| \tag{6.2}$$

$$\max_{t \in T}\left[\min\left[0, \tilde{q}(u)(t)\right] + \langle \nabla \tilde{q}(u)(t), v - u\rangle\right] \leq -K_1 \max_{i \in E}\left|\tilde{h}_i^1(u)\right| \tag{6.3}$$

$$\text{and} \quad \|v - u\|_{\mathcal{L}^2} \leq K_2 \max_{i \in E}\left|\tilde{h}_i^1(u)\right|. \tag{6.4}$$

[2] Warga considers optimal control algorithms in the space of relaxed controls. It is true that sequences of relaxed controls generated by these algorithms have accumulation points (in the relaxed topology) which satisfy the strong version of the relaxed maximum principle. However, controls generated by these algorithms are typically relaxed controls. This implies that if these algorithms perform only a finite number of iterations the calculated controls are relaxed controls. This means that we have to perform the additional step in these optimal control procedures—that of transforming a relaxed control solution to an ordinary control. Unfortunately this can be accomplished only approximately with the help of the *chattering lemma*—[11],[90].

Proof.

Take any $\tilde{\mu} \in \bar{\mathcal{U}}$. Let $r > 0$ be a number such that

$$\max_{i \in E} \left| \hat{h}_i^1(\mu) \right| < r$$

for all $\mu \in \bar{\mathcal{U}}$. We deduce from (**CQ**) that there is a simplex in $\mathcal{E}(\tilde{\mu}) \subset \mathcal{R}^{n_E}$ ($n_E = |E|$) with vertices $\{e_j\}_{j=0}^{n_E}$ which contains 0 as an interior point. By definition of $\mathcal{E}(\tilde{\mu})$, there exist $d_0, \ldots, d_{n_E} \in \mathcal{D}$ and $\delta > 0$ such that for $j = 0, \ldots, n_E$

$$\begin{aligned} \left\{ \left\langle \nabla \hat{h}_i^1(\tilde{\mu}), d_j \right\rangle_r \right\}_{i \in E} &= e_j, \\ \max_{i \in I} \left[\min\left[0, \hat{h}_i^2(\tilde{\mu})\right] + \left\langle \nabla \hat{h}_i^2(\tilde{\mu}), d_j \right\rangle_r \right] &\leq -\delta, \\ \max_{t \in T} \left[\min\left[0, \hat{q}(\tilde{\mu})(t)\right] + \left\langle \nabla \hat{q}(\tilde{\mu})(t), d_j \right\rangle_r \right] &\leq -\delta. \end{aligned}$$

Let $(\lambda_0, \lambda_1, \ldots, \lambda_{n_E})$ be the barycentric coordinates of 0 w.r.t. the vertices e_j of the simplex, i.e.

$$0 \left(= \sum_{j=0}^{n_E} \lambda_j e_j \right) = \nabla \hat{h}^1(\tilde{\mu}) \circ \sum_{j=0}^{n_E} \lambda_j d_j.$$

Here

$$\nabla \hat{h}^1(\mu) \circ d \;:=\; \left\{ \left\langle \nabla \hat{h}_i^1(\mu), d \right\rangle_r \right\}_{i \in E}.$$

We shall also write

$$\hat{h}^1(\mu) \;:=\; \left\{ \hat{h}_i^1(\mu) \right\}_{i \in E}.$$

$\nabla \tilde{h}^1(u) \circ d$ and $\tilde{h}^1(u)$ are defined analogously.

Since the vertices are in general position and 0 is an interior point, the λ_i's are all positive and we may find $\delta_1 > 0$ such that

$$\left(\lambda_0 - \sum_{j=1}^{n_E} \alpha_j, \lambda_1 + \alpha_1, \ldots, \lambda_{n_E} + \alpha_{n_E} \right)$$
$$\in \left\{ \gamma \in \mathcal{R}^{n_E+1} : \gamma_j \geq 0 \;\forall j, \sum_{j=0}^{n_E} \gamma_j = 1 \right\}$$

whenever $\alpha \in \mathcal{B}(0, \delta_1) \subset \mathcal{R}^{n_E}$. ($\mathcal{B}(0, \delta_1)$ is a ball with radius δ_1.) Furthermore, the $n_E \times n_E$ matrix $M(\mu)$ defined by

$$M(\mu)\alpha := \sum_{j=1}^{n_E} \nabla \hat{h}^1(\mu) \circ \alpha_j (d_j - d_0) \tag{6.5}$$

is invertible for $\mu = \tilde{\mu}$, from the definition of $d_j,\ j = 1, \ldots, n_E$.

In consequence of hypothesis (**CQ**) and in view of the continuity properties of the mapping $\mu \to y_r^{\mu,d}$ for fixed d we may choose a neighbourhood $\mathcal{O}(\tilde{\mu})$ of $\tilde{\mu}$ in $\bar{\mathcal{U}}$ and numbers $\hat{r} \geq r$ and $\delta_2 \in (0, \hat{r}^{-1}]$ such that for any $u \in \mathcal{U}$ satisfying $\mathbf{u} \in \mathcal{O}(\tilde{\mu})$

$$(i) \qquad \max_{i \in I}\left[\min\left[0, \tilde{h}_i^2(u)\right] + \left\langle \nabla \tilde{h}_i^2(u), v_j - u\right\rangle\right] \leq -\delta/2 \ \ \forall j$$

$$(ii) \qquad \max_{t \in T}[\min[0, \tilde{q}(u)(t)] + \langle \nabla \tilde{q}(u)(t), v_j - u\rangle] \leq -\delta/2 \ \ \forall j$$

$$(iii) \qquad M(u) \text{ is invertible}$$

$$(iv) \qquad \left\| M(u)^{-1} \nabla \tilde{h}^1(u) \circ \left(\left(\sum_{j=0}^{n_E} \lambda_j v_j \right) - u \right) \right\| \leq \delta_1/2$$

$$(v) \qquad \delta_2 \left\| M(u)^{-1} \right\| n_E^{1/2} \leq \delta_1/2.$$

(In *(v)* the norm is the Euclidean norm. $M(u)$ is defined analogously to $M(\mu)$.) Here the controls $v_j \in \mathcal{U},\ j = 0, \ldots, n_E$ are defined to be

$$v_j(t) := u(t) + d_j(t, u(t)).$$

Now suppose that the control u is feasible with respect to the inequality constraints and $\tilde{h}^1(u) \neq 0$. Set

$$\alpha = M(u)^{-1}\left[-\nabla \tilde{h}^1(u) \circ \left(\left(\sum_{j=0}^{n_E} \lambda_j v_j \right) - u \right) - \delta_2 \left\| \tilde{h}^1(u) \right\|_\infty^{-1} \tilde{h}^1(u)\right] \tag{6.6}$$

in which $\|\tilde{h}^1(u)\|_\infty := \max_{i \in E} |\tilde{h}_i^1(u)|$. Notice that, by properties *(iv)* and *(v)*, $\|\alpha\| \leq \delta_1$. Also set

$$\hat{v} = v_0 + \sum_{j=1}^{n_E} (\lambda_j + \alpha_j)(v_j - v_0).$$

Because $\|\alpha\| \leq \delta_1$ we have that $\hat{v} \in \mathcal{U}$. Finally we define v to be

$$v = u + \left(\left\| \tilde{h}^1(u) \right\|_\infty / \hat{r} \right)(\hat{v} - u).$$

Since $\|\tilde{h}^1(u)\|_\infty/\hat{r} \leq 1$, it follows that $v \in \mathcal{U}$. We now verify that this control function has the required properties.

Notice first that

$$\|v - u\|_{\mathcal{L}^2} \leq (2d/\hat{r}) \left\| \tilde{h}_i^1(u) \right\|_\infty, \tag{6.7}$$

where d is a bound on the $\mathcal{L}_m^2[T]$ norms of elements in $\mathcal{U}$.

We have from (6.5) and (6.6) that

$$\begin{aligned} M(u)\alpha &= \nabla\tilde{h}^1(u)\circ\sum_{j=1}^{n_E}\alpha_j(v_j-v_0) \\ &= -\nabla\tilde{h}^1(u)\circ\left(\sum_{j=1}^{n_E}\lambda_j\,(v_j-v_0)+v_0-u\right)-\delta_2\left\|\tilde{h}^1(u)\right\|_\infty^{-1}\tilde{h}^1(u). \end{aligned}$$

By definition of $\hat{v}$,

$$\nabla\tilde{h}^1(u)\circ(\hat{v}-u)=-\delta_2\left\|\tilde{h}^1(u)\right\|_\infty^{-1}\tilde{h}^1(u).$$

But then

$$\nabla\tilde{h}^1(u)(v-u)=-(\delta_2/\hat{r})\tilde{h}^1(u).$$

Since $\delta_2/\hat{r}\le 1$, it follows that

$$\begin{aligned} \max_{i\in E}\left|\tilde{h}_i^1(u)+\left\langle\nabla\tilde{h}_i^1(u),v-u\right\rangle\right|-\max_{i\in E}\left|\tilde{h}_i^1(u)\right| \\ \le -(\delta_2/\hat{r})\left\|\tilde{h}^1(u)\right\|_\infty. \end{aligned} \tag{6.8}$$

We deduce from property *(i)* that

$$\left\langle\nabla\tilde{h}_j^2(u),v_0+\sum_{i=1}^{n_E}(\lambda_i+\alpha_i)\,(v_i-v_0)-u\right\rangle \le -\min\left[0,\tilde{h}_j^2(u)\right]-\delta/2, \quad \forall j\in I.$$

It follows that

$$\left\langle\nabla\tilde{h}_j^2(u),v-u\right\rangle\le\left(\left\|\tilde{h}^1(u)\right\|_\infty/\hat{r}\right)\left(-\min\left[0,\tilde{h}_j^2(u)\right]-\delta/2\right)\ \forall j\in I.$$

Since $\|\tilde{h}^1(u)\|_\infty/\hat{r}\le 1$ and

$$-\min\left[0,\tilde{h}_j^2(u)\right]\ge 0\ \forall j\in I,$$

we deduce that

$$\min\left[0,\tilde{h}_j^2(u)\right]+\left\langle\nabla\tilde{h}_j^2(u),v-u\right\rangle\le-(\delta/(2\hat{r}))\left\|\tilde{h}^1(u)\right\|_\infty\ \ \forall j\in I. \tag{6.9}$$

Likewise we deduce from property *(ii)* that

$$\min\left[0,\tilde{q}(u)(t)\right]+\langle\nabla\tilde{q}(u)(t),v-u\rangle\le-(\delta/(2\hat{r}))\left\|\tilde{h}^1(u)\right\|_\infty\ \ \forall t\in T. \tag{6.10}$$

Surveying inequalities (6.7)–(6.10), we see that v satisfies all relevant conditions for completion of the proof, when we set $K_1=\min\{\delta_2,\delta/(2\hat{r})\}$ and $K_2=2d/\hat{r}$, numbers whose magnitudes do not depend on our choice of u. ∎

Proof.

(*Proposition 4.1*) *(i)* Fix $u \in \mathcal{U}$ satisfying the inequality constraints. We must find $\hat{c} > 0$ such that, if $c > \hat{c}$ then $t_c(u) \leq 0$. If $M(u) := \max_{i \in E} |\tilde{h}_i^1(u)| = 0$ then of course $t_c(u) \leq 0$ for any $c > 0$. If $M(u) > 0$ then according to *Lemma 6.1* there exists $\hat{d} \in \mathcal{U} - u$ such that, if we set $\varepsilon = K_1 M(u) > 0$ with K_1 as in *Lemma 6.1*, then

$$\theta(u) < -\varepsilon.$$

Here

$$\begin{aligned}\theta(u) = \max \Bigg[& \max_{i \in E} \left|\tilde{h}_i^1(u) + \left\langle \nabla \tilde{h}_i^1(u), \hat{d} \right\rangle\right| - \max_{i \in E} \left|\tilde{h}_i^1(u)\right|, \\ & \max_{j \in I} \left[\tilde{h}_j^2(u) + \left\langle \nabla \tilde{h}_j^2(u), \hat{d} \right\rangle\right], \\ & \max_{t \in T} \left[\tilde{q}(u)(t) + \left\langle \nabla \tilde{q}(u)(t), \hat{d} \right\rangle\right]\Bigg].\end{aligned}$$

Because $\sigma_c(u) \leq \langle \nabla \tilde{F}_0(u), \hat{d} \rangle / c + \theta(u)$, from the definition of t_c and *Proposition 2.1.1*, we get

$$t_c(u) \leq \left[W + \max_{i \in E} \left|\tilde{h}_i^1(u)\right|\right] / c + \theta(u),$$

where $W = \max\,[0, \langle \nabla \tilde{F}_0(u), \hat{d} \rangle]$. It follows that $t_c(u) \leq 0$ for any $c > \hat{c}$ where

$$\hat{c} := \frac{W + M(u)}{-\theta(u)}.$$

(ii) Take $u \in \mathcal{U}$ satisfying the inequality constraints and $c > 0$ such that $\sigma_c(u) < 0$. Let (d, β) be the solution to $\mathbf{P_c(u)}$. Since $\sigma_c(u) \neq 0$, it follows that $d \neq 0$.

We deduce from the differentiability properties of ϕ, h_i^1, h_j^2 and q and *Proposition 2.1.2* that there exists $o : [0, \infty) \to [0, \infty)$ such that $s^{-1} o(s) \to 0$ as $s \downarrow 0$ and the following three inequalities are valid for any $\alpha \in [0, 1]$:

$$\begin{aligned}\tilde{F}_c(u + \alpha d) - \tilde{F}_c(u) &\leq \alpha \left\langle \nabla \tilde{F}_0(u), d \right\rangle / c + \max_{i \in E} \left|\tilde{h}_i^1(u) + \alpha \left\langle \nabla \tilde{h}_i^1(u), d \right\rangle\right| \\ &\quad - \max_{i \in E} \left|\tilde{h}_i^1(u)\right| + o(\alpha), &(6.11)\\ \tilde{h}_j^2(u + \alpha d) &\leq \tilde{h}_j^2(u) + \alpha \left\langle \nabla \tilde{h}_j^2(u), d \right\rangle + o(\alpha) \quad \forall j \in I, &(6.12)\\ \tilde{q}(u + \alpha d)(t) &\leq \tilde{q}(u)(t) + \alpha \langle \nabla \tilde{q}(u)(t), d \rangle + o(\alpha) \quad \forall t \in T. &(6.13)\end{aligned}$$

By convexity of the function $e \to \max_{i \in E} |\tilde{h}_i^1(u) + \langle \nabla \tilde{h}_i^1(u), e \rangle|$:

$$\begin{aligned}&\max_{i \in E} \left|\tilde{h}_i^1(u) + \alpha \left\langle \nabla \tilde{h}_i^1(u), d \right\rangle\right| - \max_{i \in E} \left|\tilde{h}_i^1(u)\right| \leq \\ &\alpha \left[\max_{i \in E} \left|\tilde{h}_i^1(u) + \left\langle \nabla \tilde{h}_i^1(u), d \right\rangle\right| - \max_{i \in E} \left|\tilde{h}_i^1(u)\right|\right].\end{aligned}$$

From inequality (6.11) then

$$\begin{aligned}\tilde{F}_c(u+\alpha d)-\tilde{F}_c(u) &\leq \alpha\left[\left\langle\nabla\tilde{F}_0(u),d\right\rangle/c+\max_{i\in E}\left|\tilde{h}_i^1(u)+\left\langle\nabla\tilde{h}_i^1(u),d\right\rangle\right|\right.\\ &\qquad\left.-\max_{i\in E}\left|\tilde{h}_i^1(u)\right|\right]+o(\alpha)\\ &\leq \alpha\sigma_c(u)+o(\alpha).\end{aligned}$$

It follows that

$$\tilde{F}_c(u+\alpha d)-\tilde{F}_c(u)\leq\alpha\gamma\sigma_c(u)\ \ \forall\alpha\in[0,\alpha_1], \tag{6.14}$$

where $\alpha_1>0$ is such that $o(\beta)\leq\beta(\gamma-1)\sigma_c(u)$ for all $\beta\in[0,\alpha_1]$.

It remains to show that $u+\alpha d$ is feasible w.r.t. the inequality constraints for sufficiently small α. Since $\tilde{h}_j^2(u)\leq 0$ and $\alpha\in[0,1]$ (6.12) implies

$$\begin{aligned}\tilde{h}_j^2(u+\alpha d) &\leq \alpha\left[\tilde{h}_j^2(u)+\left\langle\nabla\tilde{h}_j^2(u),d\right\rangle\right]+o(\alpha)\\ &\leq \alpha\sigma_c(u)+o(\alpha)\leq\alpha\gamma\sigma_c(u)<0\end{aligned}$$

for all $\alpha\in[0,\alpha_1]$, as required.

We deduce from the differentiability properties of $\tilde{q}$ that

$$\begin{aligned}\max_{t\in T}\tilde{q}(u+\alpha d)(t) &\leq \max_{t\in R_{\epsilon,u}}[\tilde{q}(u)(t)+\alpha\langle\nabla\tilde{q}(u)(t),d\rangle]+o(\alpha)\\ &\leq \alpha\sigma_c(u)+o(\alpha)\leq\alpha\gamma\sigma_c(u)<0,\end{aligned}$$

for $\alpha\in[0,\alpha_1]$, as required. ∎

Proof.

(*Theorem 5.1*) *(i)* Let $\{u_k\}$ be the sequence generated by *Algorithm 1* and let $\{c_k\}$ be the corresponding penalty parameters. Let $\{k_l\}$ be the sequence of index values at which the penalty parameter increases. By extracting a further subsequence (we do not relabel) we can arrange that the sequence $\{\mathbf{u}_{k_l}\}$ has a limit point $\bar{\mu}\in\mathcal{U}$ because $\mathcal{U}$ is compact. We shall find a number $\hat{c}<\infty$ such that for sufficiently large k_l '$c_{k_l}>\hat{c}$' implies '$\sigma_{c_{k_l}}(u_{k_l})\leq-\max_{i\in E}|\tilde{h}_i^1(u_{k_l})|/c_{k_l}$'. This contradicts our assumption that the penalty parameter increases along the subsequence. So we may conclude that $\{c_k\}$ is bounded.

Fix k_l such that $\mathbf{u}_{k_l}\in\mathcal{O}(\bar{\mu})$, where $\mathcal{O}(\bar{\mu})$ is the neighbourhood of $\bar{\mu}$ as specified in *Lemma 6.1*. From the minimizing property of $\sigma_{c_{k_l}}(u_{k_l})$ we deduce

$$\begin{aligned}\sigma_{c_{k_l}}(u_{k_l}) &\leq 1/(2c_{k_l})\,\|v_{k_l}-u_{k_l}\|_{\mathcal{L}^2}^2+\left\langle\nabla\tilde{F}_0(u_{k_l}),v_{k_l}-u_{k_l}\right\rangle/c_{k_l}+\\ &\qquad\max\left[\max_{i\in E}\left|\tilde{h}_i^1(u_{k_l})+\left\langle\nabla\tilde{h}_i^1(u_{k_l}),v_{k_l}-u_{k_l}\right\rangle\right|-\right.\end{aligned}$$

$$\max_{i\in E}\left|\tilde{h}_i^2(u_{k_l})\right|, \max_{j\in I}\left[\tilde{h}_j^2(u_{k_l}) + \left\langle \nabla\tilde{h}_j^2(u_{k_l}), v_{k_l} - u_{k_l}\right\rangle\right],$$
$$\left.\max_{t\in T}\left[\tilde{q}(u_{k_l})(t) + \langle\nabla\tilde{q}(u_{k_l})(t), v_{k_l} - u_{k_l}\rangle\right]\right] \tag{6.15}$$

for a control function v_{k_l} satisfying conditions (6.1)–(6.4) of *Lemma 6.1* in which v_{k_l}, u_{k_l} replace v, u respectively. It follows

$$\begin{aligned}\sigma_{c_{k_l}}(u_{k_l}) \quad &\leq \quad 1/(2c_{k_l})\int_0^1 \|v_{k_l}(t) - u_{k_l}(t)\|^2\,dt + \\ &\left|\left\langle \nabla\tilde{F}_0(u_{k_l}), v_{k_l} - u_{k_l}\right\rangle\right|/c_{k_l} - K_1\max_{i\in E}\left|\tilde{h}_i^1(u_{k_l})\right|.\end{aligned}$$

Since the control constraint Ω is bounded and in view of *Proposition 2.1.1*, there exists $r > 0$ (independent of k_l) such that

$$\begin{aligned}1/(2c_{k_l})\int_0^1 \|v_{k_l}(t) - u_{k_l}(t)\|^2\,dt + \left|\left\langle\nabla\tilde{F}_0(u_{k_l}), v_{k_l} - u_{k_l}\right\rangle\right|/c_{k_l} \\ \leq (r/c_{k_l})\|v_{k_l} - u_{k_l}\|_{\mathcal{L}^2}.\end{aligned}$$

Hence

$$\sigma_{c_{k_l}}(u_{k_l}) \leq -(K_1 - rK_2/c_{k_l})\max_{i\in E}\left|\tilde{h}_i^1(u_{k_l})\right|.$$

We conclude that

$$\sigma_{c_{k_l}}(u_{k_l}) \leq -\max_{i\in E}\left|\tilde{h}_i^1(u_{k_l})\right|/c_{k_l}$$

if $c_{k_l} \geq \hat{c}$, where $\hat{c} = K_1^{-1}(rK_2 + 1)$.

(ii) and *(iii)* Let $\{u_k\}$ be an infinite sequence generated by the algorithm. We must show that $\lim_{k\to\infty}\sigma_{c_k}(u_k) = 0$ and, if a convergent subsequence of $\{\mathbf{u}_k\}$ has a limit point $\bar{\mu}\in\mathcal{U}$, that conditions ($\mathbf{NC^r}$) are satisfied at $\bar{\mu}$.

Stage 1. (Convergence analysis) Since the c_k's are bounded and can increase only by multiplies of c^0, we must have $c_k = c$ for all $k \geq k_0$, for some k_0 and $c > 0$. In view of the manner in which u_k's are constructed, we have

$$\tilde{F}_c(u_{k+1}) - \tilde{F}_c(u_k) \leq \gamma\alpha_k\sigma_c(u_k)$$

for all $k \geq k_0$. This means that, for all $j \geq 1$, $k \geq k_0$

$$\tilde{F}_c(u_{k+j}) - \tilde{F}_c(u_k) \leq \gamma\sum_{i=0}^{j-1}\alpha_{k+i}\sigma_c(u_{k+i}). \tag{6.16}$$

Since $\{\tilde{F}_c(u_k)\}$ is a bounded sequence and $\alpha_k\sigma_c(u_k)$ are nonpositive, we conclude

$$\alpha_k\sigma_c(u_k) \to 0 \text{ as } k\to\infty. \tag{6.17}$$

Since $\sigma_c(\mu)$ is bounded as μ ranges over $\bar{\mathcal{U}}$ we can arrange by a subsequence extraction (we do not relabel) that

$$\sigma_c(u_k) \to \beta \text{ for some } \beta \leq 0.$$

We claim that $\beta = 0$. To show this, suppose to the contrary that $\beta < 0$. Then by (6.17) $\alpha_k \to 0$.

We must have

$$\begin{aligned} \tilde{F}_c(u_k + \eta^{-1}\alpha_k d_k) - \tilde{F}_c(u_k) &> \gamma\eta^{-1}\alpha_k\sigma_c(u_k) \text{ or} \\ \max_{j\in I} \tilde{h}_j^2(u_k + \eta^{-1}\alpha_k d_k) &> 0, \text{ or} \\ \max_{t\in T} \tilde{q}(u_k + \eta^{-1}\alpha_k d_k)(t) &> 0 \end{aligned}$$

for all k sufficiently large.

However $\|\alpha_k d_k\|_{\mathcal{L}^\infty} \to 0$ as $k \to \infty$. We deduce then from *Proposition 2.1.3* and the continuity of $\tilde{q}$ that

$$\max_{t\in R_{\epsilon,u_k}} \tilde{q}(u_k + \eta^{-1}\alpha_k d_k)(t) = \max_{t\in T} \tilde{q}(u_k + \eta^{-1}\alpha_k d_k)(t)$$

for all k sufficiently large. Since $\sigma_c(u_k) \leq 0$, we conclude that

$$\begin{aligned} \max \Big[&\tilde{F}_c(u_k + \eta^{-1}\alpha_k d_k) - \tilde{F}_c(u_k), \max_{j\in I} \tilde{h}_j^2(u_k + \eta^{-1}\alpha_k d_k), \\ &\max_{t\in R_{\epsilon,u_k}} \tilde{q}(u_k + \eta^{-1}\alpha_k d_k)(t) \Big] \geq \gamma\eta^{-1}\alpha_k\sigma_c(u_k). \end{aligned} \tag{6.18}$$

It follows now from *Props. 2.1.4–2.1.6* that there exists a function $o : [0,\infty) \to [0,\infty)$ such that $s^{-1}o(s) \to 0$ as $s \downarrow 0$ and

$$\begin{aligned} \max_{j\in I} \tilde{h}_j^2(u_k + \eta^{-1}\alpha_k d_k) &\leq \max_{j\in I} \left[\tilde{h}_j^2(u_k) + \left\langle \nabla\tilde{h}_j^2(u_k), \eta^{-1}\alpha_k d_k \right\rangle \right] + \\ &\quad o(\eta^{-1}\alpha_k \|d_k\|_{\mathcal{L}^\infty}), \end{aligned} \tag{6.19}$$

$$\begin{aligned} \max_{t\in R_{\epsilon,u_k}} \tilde{q}(u_k + \eta^{-1}\alpha_k d_k)(t) &\leq \max_{t\in R_{\epsilon,u_k}} \left[\tilde{q}(u_k)(t) + \langle \nabla\tilde{q}(u_k), \eta^{-1}\alpha_k d_k \rangle \right] + \\ &\quad o(\eta^{-1}\alpha_k \|d_k\|_{\mathcal{L}^\infty}) \end{aligned} \tag{6.20}$$

and

$$\begin{aligned} \tilde{F}_c(u_k + \eta^{-1}\alpha_k d_k) - \tilde{F}_c(u_k) &\leq \left\langle \nabla\tilde{F}_0(u_k), \eta^{-1}\alpha_k d_k \right\rangle / c + \\ &\quad \max_{i\in E} \left| \tilde{h}_i^1(u_k) + \left\langle \nabla\tilde{h}_i^1(u_k), \eta^{-1}\alpha_k d_k \right\rangle \right| \\ &\quad - \max_{i\in E} \left| \tilde{h}_i^1(u_k) \right| + o(\eta^{-1}\alpha_k \|d_k\|_{\mathcal{L}^\infty}). \end{aligned} \tag{6.21}$$

Since u_k is feasible w.r.t. the inequality constraints, we have

$$\max_{j\in I}\left[\tilde{h}_j^2(u_k)+\left\langle\nabla\tilde{h}_j^2(u_k),\eta^{-1}\alpha_k d_k\right\rangle\right] \leq \eta^{-1}\alpha_k \max_{j\in I}\left[\tilde{h}_j^2(u_k)+\left\langle\nabla\tilde{h}_j^2(u_k),d_k\right\rangle\right] \quad (6.22)$$

and

$$\max_{t\in R_{\epsilon,u_k}}\left[\tilde{q}(u_k)(t)+\langle\nabla\tilde{q}(u_k)(t),\eta^{-1}\alpha_k d_k\rangle\right] \leq \eta^{-1}\alpha_k \max_{t\in R_{\epsilon,u_k}}\left[\tilde{q}(u_k)(t)+\langle\nabla\tilde{q}(u_k)(t),d_k\rangle\right] \quad (6.23)$$

for k sufficiently large. Also, by the convexity of $e \to \max_{i\in E}|\tilde{h}_i^1(u_k)+\langle\nabla\tilde{h}_i^1(u_k),e\rangle|$,

$$\max_{i\in E}\left|\tilde{h}_i^1(u_k)+\left\langle\nabla\tilde{h}_i^1(u_k),\eta^{-1}\alpha_k d_k\right\rangle\right| - \max_{i\in E}\left|\tilde{h}_i^1(u_k)\right|$$
$$\leq \eta^{-1}\alpha_k\left(\max_{i\in E}\left|\tilde{h}_i^1(u_k)+\left\langle\nabla\tilde{h}_i^1(u_k),d_k\right\rangle\right| - \max_{i\in E}\left|\tilde{h}_i^1(u_k)\right|\right). \quad (6.24)$$

Combining inequalities (6.18)–(6.24), noting the definition of $\sigma_c(u_k)$ and the fact that (d_k,β_k) solves $\mathbf{P_c(u_k)}$, and dividing across the resulting inequality by α_k we arrive at

$$\eta^{-1}\sigma_c(u_k)+\alpha_k^{-1}o(\eta^{-1}\alpha_k\|d_k\|_{\mathcal{L}^\infty})\geq\eta^{-1}\gamma\sigma_c(u_k).$$

We get $\eta^{-1}\beta\geq\eta^{-1}\gamma\beta$ in the limit. But this implies $\gamma\geq 1$, since $\beta<0$ by assumption. From this contradiction we conclude the validity of $\beta=0$. Assertion *(ii)* of the theorem follows from the definition of t_c and part *(i)*.

Let $\{\mathbf{u}_k\}$ be a convergent subsequence with the limit point $\bar{\mu}\in\bar{\mathcal{U}}$. We must show that conditions $(\mathbf{NC^r})$ are satisfied at $\bar{\mu}$. First we establish that $\bar{\mu}$ is feasible for $(\mathbf{P^r})$ and, for some $c>0$,

$$0\leq\left[\left\langle\nabla\hat{F}_0(\bar{\mu}),d\right\rangle_r/c+\max_{i\in E}\left|\left\langle\nabla\hat{h}_i^1(\bar{\mu}),d\right\rangle_r\right|,\ \max_{j\in I}\left\langle\nabla\hat{h}_j^2(\bar{\mu}),d\right\rangle_r,\ \max_{t\in R_{0,\bar{\mu}}}\langle\nabla\hat{q}(\bar{\mu})(t),d\rangle_r\right]$$

for all $d\in\mathcal{D}$.

Since $\{\mathbf{u}_k\}\to\bar{\mu}$ we know that $x_r^{\mathbf{u}_k}\to x_r^{\bar{\mu}}$ uniformly. Because u_k is feasible w.r.t. the inequality constraints, and the penalty parameter is not updated for k sufficiently large we have

$$\sigma_c(u_k)\leq-\max_{i\in E}\left|\tilde{h}_i^1(u_k)\right|/c,\ \max_{j\in I}\tilde{h}_j^2(u_k)\leq 0,\ \max_{t\in T}\tilde{q}(u_k)(t)\leq 0.$$

But we have shown that $\sigma_c(u_k) \to 0$ as $k \to \infty$. It follows now from the fact that $\mathbf{u}_k \to \bar{\mu} \in \bar{\mathcal{U}}$ that in the limit

$$\max_{i \in E} \left| \hat{h}_i^1(\bar{\mu}) \right| = 0, \ \max_{j \in I} \hat{h}_j^2(\bar{\mu}) \leq 0, \ \max_{t \in T} \hat{q}(\bar{\mu})(t) \leq 0. \tag{6.25}$$

We have established that $\bar{\mu}$ is feasible for $(\mathbf{P^r})$.

Now choose any sequence $\rho_k \downarrow 0$, $\rho_k \leq 1$ for all k such that

$$\rho_k^{-1} \sigma_c(u_k) \to 0 \quad \text{as } k \to \infty. \tag{6.26}$$

Choose any $d \in \mathcal{D}$. By the convexity of $\mathcal{U}$, $\rho_k d(\cdot, u_k) \in \mathcal{U} - u_k$ for each k.

By definition of σ_c then

$$\begin{aligned} \sigma_c(u_k) \ \leq \ & 1/2\rho_k^2 \, \|d(\cdot, u_k)\|_{\mathcal{L}^2} / c + \max \Big[\phi_x(x^{u_k}(1)) y^{u_k, \rho_k d(\cdot, u_k)}(1)/c + \\ & \max_{i \in E} \left| h_i^1(x^{u_k}(1)) + (h_i^1)_x(x^{u_k}(1)) y^{u_k, \rho_k d(\cdot, u_k)}(1) \right| \\ & - \max_{i \in E} \left| h_i^1(x^{u_k}(1)) \right|, \\ & \max_{j \in I} \left[h_j^2(x^{u_k}(1)) + (h_j^2)_x(x^{u_k}(1)) y^{u_k, \rho_k d(\cdot, u_k)}(1) \right], \\ & \max_{t \in T} \left[q(t, x^{u_k}(t)) + q_x(t, x^{u_k}(t)) y^{u_k, \rho_k d(\cdot, u_k)}(t) \right] \Big]. \end{aligned} \tag{6.27}$$

Fix $\hat{\varepsilon} > 0$. Since $\rho_k \downarrow 0$ (and consequently $y^{u_k, \rho_k d(\cdot, u_k)} \to 0$ uniformly) and also u_k is feasible w.r.t. the inequality constraints for each k, we have:

$$\begin{aligned} & \max_{j \in I} \left[h_j^2(x^{u_k}(1)) + (h_j^2)_x(x^{u_k}(1)) y^{u_k, \rho_k d(\cdot, u_k)}(1) \right] \\ & \qquad \leq \max_{I_{\hat{\varepsilon}, \bar{\mu}}} \, (h_j^2)_x(x^{u_k}(1)) y^{u_k, \rho_k d(\cdot, u_k)}(1), \\ & \max_{t \in T} \left[q(t, x^{u_k}(t)) + q_x(t, x^{u_k}(t)) y^{u_k, \rho_k d(\cdot, u_k)}(t) \right] \\ & \qquad \leq \max_{t \in R_{\hat{\varepsilon}, \bar{\mu}}} q_x(t, x^{u_k}(t)) y^{u_k, \rho_k d(\cdot, u_k)}(t) \end{aligned}$$

for all k sufficiently large. Inserting these inequalities into (6.27), noting that $y^{u_k, \rho_k d(\cdot, u_k)} = \rho_k y^{u_k, d(\cdot, u_k)}$, dividing across by ρ_k and passing to the limit with the help of (6.26) and continuity of $\hat{F}_0(\cdot)$, $\mu \to \langle \nabla \hat{F}_0(\mu, d \rangle_r$, etc, we obtain

$$\begin{aligned} 0 \leq \max \Big[& \phi_x(x_r^{\bar{\mu}}(1)) y_r^{\bar{\mu}, d}(1)/c + \max_{i \in E} \left| (h_i^1)_x(x_r^{\bar{\mu}}(1)) y_r^{\bar{\mu}, d}(1) \right|, \\ & \max_{j \in I_{\hat{\varepsilon}, \bar{\mu}}} (h_j^2)_x(x_r^{\bar{\mu}}(1)) y_r^{\bar{\mu}, d}(1), \ \max_{t \in R_{\hat{\varepsilon}, \bar{\mu}}} q_x(t, x_r^{\bar{\mu}}(t)) y_r^{\bar{\mu}, d}(t) \Big]. \end{aligned} \tag{6.28}$$

This inequality is valid for each $\hat{\varepsilon} > 0$ and $d \in \mathcal{D}$.

Again choose arbitrary $d \in \mathcal{D}$ and take $\varepsilon_k \downarrow 0$. For each k let $(h_j^2)_x(x_r^{\bar{\mu}}(1)) \cdot y_r^{\bar{\mu},d}(1)$ achieve its maximum over $I_{\varepsilon_k,\bar{\mu}}$ at $j = j_k$ and let $q_x(t, x_r^{\bar{\mu}}(t))y_r^{\bar{\mu},d}(t)$ achieve its maximum over $t \in R_{\varepsilon_k,\bar{\mu}}$ at $t = t_k$. Then

$$0 \leq \max\Big[\phi_x(x_r^{\bar{\mu}}(1))y_r^{\bar{\mu},d}(1)/c + \max_{i\in E}\big|(h_i^1)_x(x_r^{\bar{\mu}}(1))y_r^{\bar{\mu},d}(1)\big|,$$
$$(h_{j_k}^2)_x(x_r^{\bar{\mu}}(1))y_r^{\bar{\mu},d}(1),\; q_x(t_k, x_r^{\bar{\mu}}(t_k))y_r^{\bar{\mu},d}(t_k)\Big]. \qquad (6.29)$$

Extract a subsequence (we do not relabel) such that $j_k = \bar{j}$ for all k and $t_k \to \bar{t}$ for some index value $\bar{j}$ and some $\bar{t}$. By continuity of the functions involved $\bar{j} \in I_{0,\bar{\mu}}$, $\bar{t} \in R_{0,\bar{\mu}}$ and (6.29) is valid with $\bar{j}$ and $\bar{t}$ replacing j_k and t_k respectively.

We have arrived at

$$0 \leq \max\Big[\phi_x(x_r^{\bar{\mu}}(1))y_r^{\bar{\mu},d}(1)/c + \max_{i\in E}\big|(h_i^1)_x(x_r^{\bar{\mu}}(1))y_r^{\bar{\mu},d}(1)\big|,$$
$$\max_{j\in I_{0,\bar{\mu}}}(h_j^2)_x(x_r^{\bar{\mu}}(1))y_r^{\bar{\mu},d}(1),\; \max_{t\in R_{0,\bar{\mu}}} q_x(t, x_r^{\bar{\mu}}(t))y_r^{\bar{\mu},d}(t)\Big].$$

This inequality, which holds for all $d \in \mathcal{D}$, in particular for $(d \equiv 0) \in \mathcal{D}$, is what we set out to prove.

Finally, we must attend to the case when *Algorithm 1* generates a finite sequence which terminates at a control $\mathbf{u}_{\bar{k}} = \bar{\mu}$ satisfying the stopping criterion. This case is dealt with by applying the preceding arguments to the infinite sequence of controls obtained by 'filling in' with repetitions of the following control u_k.

Stage 2. (Dualization) . The conclusions of Stage 1 can be expressed

$$\min_{d\in\mathcal{D}}\max_{\gamma\in\mathcal{K}}\Phi(d,\gamma) = 0$$

where

$$\mathcal{K} := \Big\{\gamma = (\alpha_0, \{\alpha_i^1\}_{i\in E}, \{\alpha_j^2\}_{j\in I}, \nu) \in \mathcal{R}^{1+|E|+|I|} \times C^*(T) :$$
$$\alpha_0 \geq 0,\; \alpha_j^2 \geq 0,\; j \in I,\; \sum_{i\in E}|\alpha_i^1| \leq \alpha_0,\; \alpha_0 + \sum_{j\in I}\alpha_j^2 + \int_T \nu(dt) = 1,$$
$$\alpha_j^2 = 0 \;\text{ if } j \notin I_{0,\bar{\mu}},\; \nu \geq 0,\; \mathrm{supp}\{\nu\} \subset R_{0,\bar{\mu}}\;\Big\}$$

and

$$\Phi(d,\gamma) := \alpha_0\left\langle\nabla\hat{F}_0(\bar{\mu}), d\right\rangle_r/c + \sum_{i\in E}\alpha_i^1\left\langle\nabla\hat{h}_i^1(\bar{\mu}), d\right\rangle_r$$
$$+ \sum_{j\in I_{0,\bar{\mu}}}\alpha_j^2\left\langle\nabla\hat{h}_j^2(\bar{\mu}), d\right\rangle_r + \int_T\langle\nabla\hat{q}(\bar{\mu})(t), d\rangle_r\,\nu(dt).$$

$\Phi(\cdot,\gamma)$ is a linear function on $\mathcal{L}^1(T,\mathcal{C}(\Omega))$ of which $\mathcal{D}$ is a convex subset. $\Phi(d,\cdot)$ is a bounded linear map and $\mathcal{K}$ is a compact convex set with respect to the product topology of $\mathcal{R}^{1+|E|+|I|}\times\mathcal{C}^*(T)$, in which the weak star topology is imposed on the last component. It follows from the minimax theorem ([3]) that there exists some nonzero $\bar{\gamma}\in\mathcal{K}$ such that

$$\min_{d\in\mathcal{D}}\max_{\gamma\in\mathcal{K}}\Phi(d,\gamma)=\min_{d\in\mathcal{D}}\Phi(d,\bar{\gamma})=0, \tag{6.30}$$

with $\bar{\gamma}=(\bar{\alpha}_0,\{\bar{\alpha}_i^1\}_{i\in E},\{\bar{\alpha}_j^2\}_{j\in I},\bar{\nu})$.

We readily deduce from the constraint qualification (**CQ**) that $\bar{\alpha}_0\neq 0$. By scaling the multipliers we may arrange that $\bar{\alpha}_0/c=1$.

Now define p_r to be the solution to the differential equation

$$-\dot{p}_r(t)=(f_x)_r(t,x_r^{\bar{\mu}}(t),\bar{\mu}(t))^T\left(p_r(t)+\int_{[0,t)}q_x(s,x_r^{\bar{\mu}}(s))^T\bar{\nu}(ds)\right)$$

and

$$-p_r(1)^T=\int_{[0,1]}q_x(s,x_r^{\bar{\mu}}(s))\bar{\nu}(ds)+\phi_x(x_r^{\bar{\mu}}(1))+$$
$$\sum_{i\in E}\bar{\alpha}_i^1(h_i^1)_x(x_r^{\bar{\mu}}(1))+\sum_{j\in I}\bar{\alpha}_j^2(h_j^2)_x(x_r^{\bar{\mu}}(1)).$$

We have, for any $d\in\mathcal{D}$,

$$\Phi(d,\bar{\gamma})+\int_{[0,1]}\left(p_r(t)^T+\int_{[0,t)}q_x(s,x_r^{\bar{\mu}}(s))\bar{\nu}(ds)\right)(\dot{y}_r^{\bar{\mu},d}(t)-$$
$$(f_x)_r(t,x_r^{\bar{\mu}}(t),\bar{\mu}(t))y_r^{\bar{\mu},d}(t)-\int_\Omega f_u(t,x_r^{\bar{\mu}}(t),u)d(t,u)\bar{\mu}(t)(du)\Big)\,dt\geq 0.$$

This inequality reduces, via an integration by parts, to

$$\int_{[0,1]}\left(p_r(t)^T+\int_{[0,t)}q_x(s,x_r^{\bar{\mu}}(s))\bar{\nu}(ds)\right)\cdot$$
$$\int_\Omega f_u(t,x_r^{\bar{\mu}}(t),u)d(t,u)\bar{\mu}(t)(du)dt\leq 0.$$

The above relationships imply that $(x_r^{\bar{\mu}},\bar{\mu})$ satisfies the stated necessary conditions. ■

7 Concluding Remarks

We have presented a general algorithm for an optimal control problem with state constraints. A notable feature of the algorithm is the fact that only

ε–active state constraints are used in the direction finding subproblems. This contrasts with other function space algorithms which include all state constraints in search directions. In the next chapter we show how the set of ε–active state constraints can be approximated by a finite number of constraints. Several numerical examples are also presented demonstrating the efficiency of our approach.

4

Implementation

We complicate our algorithm but we also make it implementable.

1 Implementable Algorithm

The algorithm presented in Chapter 3 has the direction finding subproblem $\mathbf{P_c(u)}$:

$$\min_{d \in \mathcal{U}-u, \beta \in \mathcal{R}} \beta + 1/(2c) \, ||d||^2_{\mathcal{L}^2}$$

subject to

$$\begin{aligned} \langle \nabla \tilde{F}_0(u), d\rangle / c + \max_{i \in E} \left| \tilde{h}^1_i(u) + \langle \nabla \tilde{h}^1_i(u), d\rangle \right| - \max_{i \in E} \left| \tilde{h}^1_i(u) \right| &\leq \beta \\ \tilde{h}^2_j(u) + \langle \nabla \tilde{h}^2_j(u), d\rangle &\leq \beta, \\ &\quad \forall j \in I \\ \tilde{q}(u)(t) + \langle \nabla \tilde{q}(u)(t), d\rangle &\leq \beta, \\ &\quad \forall t \in R_{\varepsilon,u}. \end{aligned} \tag{1.1}$$

Direct implementation of the algorithm is often not practical because of the large number of constraints to which (1.1) gives rise, following discretization. This difficulty is overcome by approximating the set $R_{\varepsilon,u}$ by a finite subset. As already remarked in the previous chapter, an important feature of feasible directions algorithms such as the one proposed in Chapter 3 is that adequate approximation is achievable by finite subsets containing relatively few points.

It is convenient to introduce some notation to describe the approximation of a set $R_{\varepsilon,u}$ corresponding to some $\varepsilon > 0$, $u \in \mathcal{U}$:

$$\mathcal{A}^{\xi}_{\varepsilon,u} = \{\text{finite sets } A \subset T : \; A \subset R_{\varepsilon,u} \subset A + \xi[-1,1]\}, \; \xi > 0$$

($A^{\xi}_{\varepsilon,u}$ is the set of finite inner approximations A to $R_{\varepsilon,u}$ with the property that if $t \in R_{\varepsilon,u}$ then there exists $a \in A \in \mathcal{A}^{\xi}_{\varepsilon,u}$ such that $|a - t| \leq \xi$).

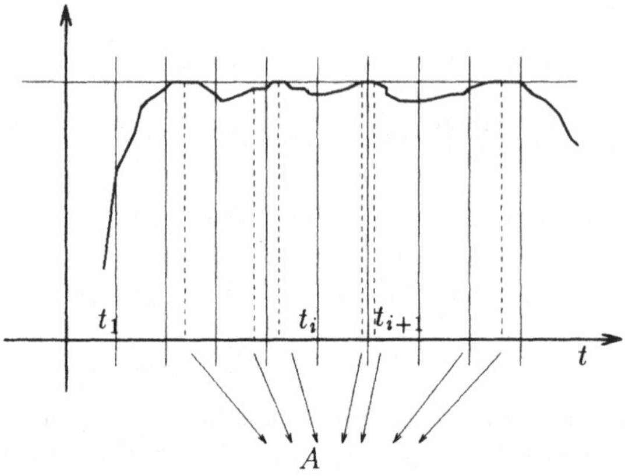

FIGURE 4.1. Approximation of state constraint.

We formulate the modified algorithm in terms of arbitrary sets drawn from $\mathcal{A}^{\xi}_{\varepsilon,u}$ to retain flexibility. The following is often found to be an effective choice of $A \in \mathcal{A}^{\xi}_{\varepsilon,u}$ however. Let $\{t_0 = 0, t_1, \ldots, t_N = 1\}$ be a uniform partition of $[0,1]$ with $|t_{i+1} - t_i| \leq \xi$. For index values $i = 0, \ldots, N-1$ such that $[t_i, t_{i+1}] \bigcap R_{\varepsilon,u} \neq \emptyset$ include in A a point on which $\tilde{q}(u)(t)$ is maximized over this set. Notice that $\xi^{-1} + 1$ provides an upper bound on the number of points in A. We call this discretization $(R_{\varepsilon,u}, \xi)$*-uniform approximation*.

Figure 4.1 illustrates the approximation of $R_{\varepsilon,u}$. As we can see from the figure extra care must be taken when $\text{argmin}\{\tilde{q}(u)(t) : t \in [t_{i-1}, t_{i+1}]\}$ is close to a point t_i of the uniform partition of $[0,1]$. It is recommended to construct A in such a way that at least one point from each $[t_i, t_{i+1}]$ is included in A.

Consider the following hypothesis:

(H3): There exists a constant $L < \infty$ such that $\|q_x(t', y) - q_x(t, x)\| \leq L(\|y - x\| + |t' - t|)$ for all $x, y \in \mathcal{R}^n$, $t', t \in T$.

We note:

Proposition 1.1 *Assume that the data for* **(P)** *satisfies hypotheses* **(H1)**, **(H2)** *and* **(H3)**. *Then corresponding to each $L > 0$ there exists $0 < K < \infty$ with the following property: for any $u \in \mathcal{U}$, any search direction d such that $\|d\|_{\mathcal{L}^\infty} \leq L$, $\varepsilon > 0$, $\xi > 0$ and $A \in \mathcal{A}^{\xi}_{\varepsilon,u}$:*

$$\max_{t \in R_{\varepsilon,u}} [\tilde{q}(u)(t) + \langle \nabla \tilde{q}(t), d \rangle] \leq \max_{t \in A} [\tilde{q}(u)(t) + \langle \nabla \tilde{q}(t), d \rangle] + K\xi.$$

Proof.

The assertions are simply proved using the fact that, under the hypotheses, $\tilde{q}(u)(\cdot)$ and $\langle \nabla \tilde{q}(u)(\cdot), d \rangle$ are Lipschitz continuous with Lipschitz constant independent of u and d. ■

Proposition 1.1 implies that if, for some $\delta > 0$,

$$\tilde{q}(u)(t) + \langle \nabla \tilde{q}(u)(t), d \rangle < -\delta$$

for all $t \in A$, then

$$\tilde{q}(u)(t) + \langle \nabla \tilde{q}(u)(t), d \rangle < 0$$

for all $t \in R_{\varepsilon,u}$, provided $\xi < \delta/K$ and $A \in \mathcal{A}^{\xi}_{\varepsilon,u}$, i.e., feasibility of this linearized inequality constraint w.r.t. the full ε–active set $R_{\varepsilon,u}$ is ensured if we have feasibility only w.r.t. a finite set A, and the greater δ the smaller the number of points which need to be included in A. Since the algorithm generates 'strictly feasible' search directions, there is scope then for a finite approximation of the ε–active set $R_{\varepsilon,u}$.

The following definitions are also needed to describe a new direction finding subproblem. The first is of an approximation $\mathcal{U}_{\omega,u}$ to the set of control functions, translated by u. We take

$$\mathcal{U}_{\omega,u} := \left\{ d \in \mathcal{L}^2_m[T] : \ d(t) \in \Omega_{\omega,u(t)} \text{ a.e. on } T \right\}$$

where

$$\Omega_{\omega,u(t)} := \Big\{ d \in \mathcal{R}^m : \begin{array}{ll} (u(t)+d)_i \geq b^i_-, & \text{if } (u(t))_i \leq b^i_- + \omega, \\ (u(t)+d)_i \leq b^i_+, & \text{if } (u(t))_i \geq b^i_+ - \omega \end{array} \Big\}.$$

Here $\omega > 0$ is a parameter.

The second is the projection of a function $u \in \mathcal{L}^2_m[T]$ onto the set of control functions defined as follows

$$P_{\mathcal{U}}[u] := \hat{u} \text{ where } (\hat{u}(t))_i = \begin{cases} b^i_-, & \text{if } (u(t))_i < b^i_- \\ (u(t))_i, & \text{if } b^i_- \leq (u(t))_i \leq b^i_+ \\ b^i_+, & \text{if } (u(t))_i > b^i_+ \end{cases}$$

Fix $\omega > 0$. For given $c > 0$, $A \subset [0,1]$ and $u \in \mathcal{U}$ denote by $\mathbf{P_{c,A}(u)}$ the modification of the direction finding subproblem $\mathbf{P_c(u)}$ in which A replaces $R_{\varepsilon,u}$ and $\mathcal{U}_{\omega,u}$ the set $\mathcal{U} - u$, namely

$$\min_{d \in \mathcal{U}_{\omega,u}, \beta \in \mathcal{R}} \beta + 1/(2c) \, \|d\|^2_{\mathcal{L}^2}$$

subject to

$$\begin{aligned} \left\langle \nabla \tilde{F}_0(u), d \right\rangle / c + \max_{i \in E} \left| \tilde{h}^1_i(u) + \left\langle \nabla \tilde{h}^1_i(u), d \right\rangle \right| - \max_{i \in E} \left| \tilde{h}^1_i(u) \right| &\leq \beta \\ \tilde{h}^2_j(u) + \left\langle \nabla \tilde{h}^2_j(u), d \right\rangle &\leq \beta \ \forall j \in I \\ \tilde{q}(u)(t) + \langle \nabla \tilde{q}(u)(t), d \rangle &\leq \beta \ \forall t \in A. \end{aligned}$$

$\mathbf{P}_{\mathbf{c},\mathbf{A}}(\mathbf{u})$ has a unique solution $(\bar{d},\bar{\beta})$ (which depends on c, A and u). We define

$$\begin{aligned}\sigma_{c,A}(u) &= \bar{\beta},\\ t_{c,A}(u) &= \bar{\beta} + \max_{i\in E}\left|\tilde{h}_i^1(u)\right|/c.\end{aligned}$$

The advantage of using the subproblem $\mathbf{P}_{\mathbf{c},\mathbf{A}}(\mathbf{u})$ (in place of $\mathbf{P}_{\mathbf{c}}(\mathbf{u})$) is that the number of inequality constraints for $\mathbf{P}_{\mathbf{c},\mathbf{A}}(\mathbf{u})$ associated with the state constraint depends on the number of points in A, which in the operation of *Algorithm 2* below incorporating $\mathbf{P}_{\mathbf{c},\mathbf{A}}(\mathbf{u})$, is typically small (see the examples in §4.3). This means that solutions to $\mathbf{P}_{\mathbf{c},\mathbf{A}}(\mathbf{u})$ can be computed by means of proximity algorithms of Mayne and Polak ([75]) with the modifications proposed by Hauser ([59]), algorithms which are known to perform most efficiently in the presence of only a moderate number of inequality constraints. Note that proximity algorithms do not require a finite-dimensional parametrization of d (at least conceptually).

Algorithm 2 Fix parameters: $\varepsilon > 0$, γ, $\eta \in (0,1)$, $0 < c^0 < \infty$, $\kappa > 1$, $\tau^0 > 0$, $\xi^0 > 0$, $\omega > 0$.

1. Choose the initial control $u_0 \in \mathcal{U}$ which satisfies $\tilde{h}_j^2(u_0) \leq 0\ \forall j \in I$ and $\tilde{q}(u_0)(t) \leq 0\ \forall t \in T$. Set $k=0$, $c_{-1} = c^0$, $\xi_0 = \xi^0$, $\tau_0 = \tau^0$.

2. Set $j=1$. Let $\tau_k^j = \tau_k$, $\xi_k^j = \xi_k$.

3. Choose $A_k^j \in \mathcal{A}_{\varepsilon,u_k}^{\xi_k^j}$. Let c_k^j be the smallest number chosen from $\{c_{k-1}, \kappa c_{k-1}, \kappa^2 c_{k-1},\ldots\}$ such that the solution (d_k^j,β_k^j) to the problem $\mathbf{P}_{\mathbf{c}_k^j,\mathbf{A}_k^j}(\mathbf{u_k})$ satisfies

$$t_{c_k^j,A_k^j}(u_k) \leq 0. \tag{1.2}$$

If $\sigma_{c_k^j,A_k^j}(u_k) = 0$ then STOP.

4. Let α_k^j be the largest number from the finite set $\{\alpha \in \{1,\eta,\eta^2,\ldots\} : \alpha > \tau_k^j\}$ such that

$$u_{k+1}^j = P_{\mathcal{U}}\left[u_k + \alpha_k^j d_k^j\right]$$

satisfies the relations

$$\begin{aligned}\tilde{F}_{c_k^j}(u_{k+1}^j) - \tilde{F}_{c_k^j}(u_k) &\leq \gamma\alpha_k^j\sigma_{c_k^j,A_k^j}(u_k) &(1.3)\\ \tilde{h}_l^2(u_{k+1}^j) &\leq 0\ \forall l \in I &(1.4)\\ \tilde{q}(u_{k+1}^j)(t) &\leq 0\ \forall t \in T. &(1.5)\end{aligned}$$

If no such α_k^j exists, let $\tau_k^{j+1} = 0.5\tau_k^j$, $\xi_k^{j+1} = 0.5\xi_k^j$, increase j by one and return to Step 3.
Otherwise let $\alpha_k = \alpha_k^j$, $d_k = d_k^j$, $c_k = c_k^j$, $\tau_k = \tau_k^j$, $A_k = A_k^j$ and $u_{k+1} = u_{k+1}^j$. Set $\xi_{k+1} = \xi_k$, $\tau_{k+1} = \tau_k$ if $\xi_k^j = \xi_k$; $\xi_{k+1} = 0.5\xi_k$, $\tau_{k+1} = 0.5\tau_k$ otherwise. Increase k by one, go to Step 2.

The role of the penalty parameter test function t_{c_k,A_k} is the same as the role of the function t_c in *Algorithm 1*, i.e., to ensure that the penalty parameter is large enough in the limit (but finite) to force satisfaction of the equality endpoint constraints. The algorithm ensures that at the kth iteration:

$$t_{c_k,A_k}(u_k) = \sigma_{c_k,A_k}(u_k) + \max_{i\in E}\left|\tilde{h}_i^1(u_k)\right| / c_k \leq 0.$$

We will show that $\sigma_{c_k,A_k}(u_k) \to 0$. Moreover, $\{c_k\}$ is bounded which together with the above inequality imply that $\lim_{k\to\infty} \max_{i\in E} |\tilde{h}_i^1(u_k)| = 0$, i.e. the limiting control (if it exists) satisfies the endpoint equality constraints. Furthermore, we will show that $\sigma_{c_k,A_k}(u_k) \to 0$ implies $\sigma_{c_k}(u_k) \to 0$.

The direction finding subproblem $\mathbf{P_{c,A}(u)}$ involves a control set $\mathcal{U}_{\omega,u}$ which is unbounded. Much use will be made in the analysis to follow of the fact that, nonetheless, solutions to $\mathbf{P_{c,A}(u)}$ are uniformly bounded. The presence of the $\|d\|^2_{\mathcal{L}^2}$ term in the cost of $\mathbf{P_{c,A}(u)}$ is crucial in establishing this result. The following lemma provides a more general result for the subproblem $\mathbf{P_{c,R_{\varepsilon,u}}(u)}$. (Note that $A \subset R_{\varepsilon,u}$.)

Lemma 1.1 *Assume* **(H1)** *and* **(H2)** *are satisfied. Given* $c > 0$ *there is a finite number* K_c *with the following property: for any* $u \in \mathcal{U}$ *and any solution* $(\bar{d},\bar{\beta})$ *to* $\mathbf{P_{c,R_{\varepsilon,u}}(u)}$ *we have*

$$\|d\|_{\mathcal{L}^\infty} \leq K_c.$$

Proof.

Fix $c > 0$. Let $(\bar{d},\bar{\beta})$ be the solution to $\mathbf{P_{c,R_{\varepsilon,u}}(u)}$. This means that $(\bar{d},(z \equiv \bar{\beta}, \bar{y} = y^{u,\bar{d}}))$ is the minimizer for the optimal control problem

$$\min_d \left[1/(2c)\int_0^1 \|d(t)\|^2\,dt + z(1)\right]$$

$$\begin{aligned}
\text{s.t. } \; & \dot{y}(t) = f_x(t,x^u(t),u(t))y(t) + f_u(t,x^u(t),u(t))d(t), \;\; \dot{z}(t) = 0,\\
& d(t) \in \Omega_{\omega,u(t)} \text{ a.e. on } T,\\
& q(t,x^u(t)) + q_x(t,x^u(t))y(t) - z(t) \leq 0 \;\; \forall t \in R_{\varepsilon,u},\\
& \text{and } z(1) \geq \max[\phi_x(x^u(1))y(1)/c+\\
& \max_{i\in E}\left|h_i^1(x^u(1)) + (h_i^1)_x(x^u(1))y(1)\right| - \max_{i\in E}\left|h_i^1(x^u(1))\right|,\\
& \max_{j\in I}\left[h_j^2(x^u(1)) + (h_j^2)_x(x^u(1))y(1)\right]\Big].
\end{aligned}$$

The following information can be deduced from the Maximum Principle ([32]): there exist nonnegative numbers α_0, α_j^2, $j \in I$, numbers α_i^1, $i \in E$, a regular nonnegative measure ν and a Lipschitz continuous function p such that

$$\alpha_0 + \int_{[0,1]} \nu(ds) + \sum_{j\in I} \alpha_j^2 = 1, \sum_{i\in E} |\alpha_i^1| \leq \alpha_0, \tag{1.6}$$

$$-\dot{p}(t) = f_x(t, x^u(t), u(t))^T \left(p(t) + \int_{[0,t)} q_x(s, x^u(s))^T \nu(ds) \right), \tag{1.7}$$

$$-\left(p(1)^T + \int_{[0,1]} q_x(s, x^u(s))\nu(ds) \right) =$$

$$\alpha_0 \phi_x(x^u(1))/c + \sum_{i\in E} \alpha_i^1 (h_i^1)_x(x^u(1)) + \sum_{j\in I} \alpha_j^2 (h_j^2)_x(x^u(1)), \tag{1.8}$$

$$d \to \left(p(t)^T + \int_{[0,t)} q_x(s, x^u(s))\nu(ds) \right) f_u(t, x^u(t), u(t))d - 1/(2c)\,\|d\|^2$$

$$\text{is maximized over } \Omega_{\omega,u(t)} \text{ at } d = \bar{d}(t) \text{ a.e. on } T. \tag{1.9}$$

It follows from (1.6)–(1.8) and Gronwall's Lemma that there is a bound on the $\mathcal{L}^\infty$ norm of p which is independent of u. In view of (1.6), there is then a bound (write it $1/2K$) on ess $\sup_{t\in T} \|p(t) + \int_{[0,t)} q_x(s, x^u(s))^T \nu(ds)\| \cdot \|f_u(t, x^u(t), u(t))\|$, independent of $u \in \mathcal{U}$ (and c). But since $0 \in \mathcal{U}_{\omega,u}$, condition (1.9) gives

$$0 \leq 1/2K \left\|\bar{d}(t)\right\| - 1/(2c) \left\|\bar{d}(t)\right\|^2 \text{ a.e. on } T,$$

i.e., $\|\bar{d}\|_{\mathcal{L}^\infty} \leq cK = K_c$. This is what we set out to prove. ∎

Before presenting our convergence result we need to clarify two points about the algorithm. They are to guarantee that Step 3 and 4 of *Algorithm 2* can always be carried out. These gaps are filled by the following proposition.

Proposition 1.2 *Assume that hypotheses* (**H1**), (**H2**), (**H3**) *and* (**CQ**) *are satisfied. Then for any $u_k \in \mathcal{U}$ satisfying the endpoint and pathwise inequality constraints of* (**P**)

(i) if $M(u_k) = 0$, then either there exist a finite l and $\alpha_k^l > \tau_k^l > 0$, $c_k^l = c_{k-1}$ such that (1.2)–(1.5) are satisfied with $j = l$, or $\sigma_{c_{k-1},A}(u_k) = 0$ (where $(d_k, (\beta_k = \sigma_{c_{k-1},A}(u_k)))$ is the solution to the problem $\mathbf{P_{c_{k-1},A}(u_k)}$ with $A = R_{\varepsilon,u}$) and Algorithm 2 cycles between Steps 3–4,

(ii) if $M(u_k) > 0$, then there exist a finite l, c_k^l and $\alpha_k^l > \tau_k^l > 0$ such that (1.2)–(1.5) are satisfied with $j = l$.

Proof.

The proof of the proposition can be carried out along the lines of the proof of *Proposition 3.4.1* if we take into account *Proposition 1.1* and *Lemma 1.1*.

First, we will show that for any $j \geq 1$ there exists $\hat{c} > 0$ such that if $c > \hat{c}$ then $t_{c,A_k^j}(u_k) \leq 0$. If $M(u_k) = 0$ then of course $t_{c,A_k^j}(u_k) \leq 0$ for any $c > 0$ and $j \geq 1$ because $\sigma_{c,A_k^j}(u_k) \leq 0$. If $M(u_k) > 0$ then according to *Lemma 3.6.1* there exists $\hat{d}_k \in \mathcal{U} - u_k$ and $\varepsilon = K_1 M(u_k) > 0$, with K_1 as in *Lemma 3.6.1*, such that

$$\theta^j(u_k) < -\varepsilon.$$

Here

$$\theta^j(u_k) = \max\left[\max_{i\in E}\left|\tilde{h}_i^1(u_k) + \langle \nabla\tilde{h}_i^1(u_k), \hat{d}_k\rangle\right| - \max_{i\in E}\left|\tilde{h}_i^1(u_k)\right|,\right.$$
$$\max_{l\in I}\left[\tilde{h}_l^2(u_k) + \left\langle \nabla\tilde{h}_l^2(u_k), \hat{d}_k\right\rangle\right],$$
$$\left.\max_{t\in A_k^j}\left[\tilde{q}(u_k)(t) + \left\langle \nabla\tilde{q}(u_k)(t), \hat{d}_k\right\rangle\right]\right].$$

Because $\sigma_{c,A_k^j}(u_k) \leq \langle \nabla\tilde{F}_0(u_k), \hat{d}_k\rangle/c + \theta^j(u_k)$, from the definition of $t_{c,A}$ and *Proposition 2.1.1*, we get

$$t_{c,A_k^j}(u_k) \leq \left[W + \max_{i\in E}\left|\tilde{h}_i^1(u_k)\right|\right]/c + \theta^j(u_k),$$

where $W = \max\,[0, \langle \nabla\tilde{F}_0(u_k), \hat{d}_k\rangle]$. It follows that $t_{c,A_k^j}(u_k) \leq 0$ for any $c > \hat{c}$, where

$$\hat{c} = \frac{W + M(u_k)}{-\varepsilon}.$$

Assume that u_k is such that $\sigma_{c_{k-1},A}(u_k) = 0$, $M(u_k) = 0$ and $A = R_{\varepsilon,u_k}$. Then, it is straightforward to show that $\sigma_{c_{k-1},A_k^j}(u_k) \to 0$ when $j \to \infty$.

Otherwise, from *Lemma 3.6.1* and *Proposition 1.1*, we can show that

$$\sigma_{c_k^j,A_k^j}(u_k) \to \bar{\sigma} < 0 \quad \text{and} \quad \|d_k^j\| \to \|\bar{d}_k\| \neq 0 \tag{1.10}$$

when $j \to \infty$.

We deduce from the differentiability properties of ϕ, h_i^1, h_j^2, q and *Proposition 2.1.2* that, for a given j, there exists $o : [0,\infty) \to [0,\infty)$ such that $s^{-1}o(s) \to 0$ as $s \downarrow 0$ and the following three inequalities are valid for any $\alpha \in [0,1]$:

$$\tilde{F}_{c_k^j}(u_k + \alpha d_k^j) - \tilde{F}_{c_k^j}(u_k) \leq \alpha\left\langle \nabla\tilde{F}_0(u_k), d_k^j\right\rangle/c_k^j +$$

$$\max_{i\in E}\left|\tilde{h}_i^1(u_k)+\alpha\left\langle\nabla\tilde{h}_i^1(u_k),d_k^j\right\rangle\right|-\max_{i\in E}\left|\tilde{h}_i^1(u_k)\right|+o(\alpha), \quad (1.11)$$

$$\tilde{h}_l^2(u_k+\alpha d_k^j)\leq\tilde{h}_l^2(u_k)+\alpha\left\langle\nabla\tilde{h}_l^2(u_k),d_k^j\right\rangle+o(\alpha)\ \forall l\in I, \quad (1.12)$$

$$\tilde{q}(u_k+\alpha d_k^j)(t)\leq\tilde{q}(u_k)(t)+\alpha\left\langle\nabla\tilde{q}(u_k)(t),d_k^j\right\rangle+o(\alpha)\ \forall t\in T. \quad (1.13)$$

By the convexity of the function $e\rightarrow\max_{i\in E}|\tilde{h}_i^1(u)+\langle\nabla\tilde{h}_i^1(u),e\rangle|$

$$\max_{i\in E}\left|\tilde{h}_i^1(u_k)+\alpha\left\langle\nabla\tilde{h}_i^1(u_k),d_k^j\right\rangle\right|-\max_{i\in E}\left|\tilde{h}_i^1(u_k)\right|\leq$$
$$\alpha\left[\max_{i\in E}\left|\tilde{h}_i^1(u_k)+\left\langle\nabla\tilde{h}_i^1(u_k),d_k^j\right\rangle\right|-\max_{i\in E}\left|\tilde{h}_i^1(u_k)\right|\right].$$

From inequality (1.11) then

$$\begin{aligned}\tilde{F}_{c_k^j}(u_k+\alpha d_k^j)-\tilde{F}_{c_k^j}(u_k) &\leq \alpha\Big[\left\langle\nabla\tilde{F}_0(u_k),d_k^j\right\rangle/c_k^j+\\ &\quad\max_{i\in E}\left|\tilde{h}_i^1(u_k)+\left\langle\nabla\tilde{h}_i^1(u_k),d_k^j\right\rangle\right|\\ &\quad-\max_{i\in E}\left|\tilde{h}_i^1(u_k)\right|\Big]+o(\alpha)\\ &\leq \alpha\sigma_{c_k^j,A_k^j}(u_k)+o(\alpha).\end{aligned}$$

It follows, from (1.10), that

$$\tilde{F}_{c_k^j}(u_k+\alpha d_k^j)-\tilde{F}_{c_k^j}(u_k)\leq\alpha\gamma\sigma_{c_k^j,A_k^j}(u_k)\ \forall\alpha\in[0,\alpha_1], \quad (1.14)$$

for j sufficiently large, where $\alpha_1>0$ is such that $o(\beta)\leq 0.5\beta(\gamma-1)\bar{\sigma}$ for all $\beta\in[0,\alpha_1]$.

Next we show that $u_k+\alpha d_k^j$ is feasible w.r.t. the inequality constraints for sufficiently small α. Since $\tilde{h}_j^2(u_k)\leq 0$ and $\alpha\in[0,1]$ (1.12) implies

$$\begin{aligned}\tilde{h}_l^2(u_k+\alpha d_k^j) &\leq \alpha\left[\tilde{h}_l^2(u_k)+\left\langle\nabla\tilde{h}_l^2(u_k),d_k^j\right\rangle\right]+o(\alpha)\\ &\leq \alpha\sigma_{c_k^j,A_k^j}(u_k)+o(\alpha)\leq\alpha\gamma\sigma_{c_k^j,A_k^j}(u_k)<0\end{aligned}$$

for all $\alpha\in[0,\alpha_1]$, as required.

We deduce from the differentiability properties of $\tilde{q}$ and *Proposition 1.1* that

$$\begin{aligned}\max_{t\in T}\tilde{q}(u_k+\alpha d_k^j)(t) &\leq \max_{t\in R_{\epsilon,u_k}}\left[\tilde{q}(u_k)(t)+\alpha\left\langle\nabla\tilde{q}(u_k)(t),d_k^j\right\rangle\right]+o(\alpha)\\ &\leq \max_{t\in A_k^j}\left[\tilde{q}(u_k)(t)+\alpha\left\langle\nabla\tilde{q}(u_k)(t),d_k^j\right\rangle\right]+K\xi_k^j+o(\alpha)\\ &\leq \alpha\sigma_{c_k^j,A_k^j}(u_k)+K\xi_k^j+o(\alpha)\leq 0.5\alpha\gamma\bar{\sigma}+K\xi_k^j<0,\end{aligned}$$

for $\alpha \in [0, \alpha_1]$ and sufficiently large j. (K in the above inequalities is as in *Proposition 1.1.*)

To complete the proof we have to take account of the fact that controls generated at iteration k are now projected onto $\mathcal{U}_{\omega,u_k}$. Notice

$$0 \in \Omega_{\omega,u_k(t)} \text{ and } \Omega_{\omega,u_k(t)} \bigcap \mathcal{B}^\infty(0,\omega) = (\Omega - u_k(t)) \bigcap \mathcal{B}^\infty(0,\omega)$$

where $\mathcal{B}^\infty(0,\omega) = \{v \in \mathcal{R}^m : \max_i |v^i| \leq \omega\}$. Since $\Omega_{\omega,u_k(t)}$ is convex and contains the origin, $(\varepsilon/\|d_k^j\|_{\mathcal{L}^\infty})d_k^j \in \mathcal{U} - u_k$ for $\varepsilon \in [0,\omega)$. This means $u_k + \alpha d_k^j \in \mathcal{U}$ when $\alpha \in [0,\bar{\alpha}]$ and j is large enough. Here $\bar{\alpha} = \omega/(2\|\bar{d}_k\|_{\mathcal{L}^\infty})$. It follows that for α sufficiently small $P_{\mathcal{U}}[u_k + \alpha d_k^j] = u_k + \alpha d_k^j$. ■

We may adapt the proof of *Theorem 3.5.1*, in the light of *Proposition 1.2*, to establish the convergence result expressed in the theorem below. Results are given in relation to necessary conditions (**NC**), in a normal form, for a control function $\bar{u}$ which is feasible for (**P**), to be a minimizer.

(**NC**): There exist nonnegative numbers α_j^2, $j \in I$, numbers α_i^1, $i \in E$, an absolutely continuous function $p : [0,1] \to \mathcal{R}^n$ and a nonnegative regular measure ν on the Borel subsets of $[0,1]$ such that

$$-\dot{p}(t) = f_x(t, x^{\bar{u}}(t), \bar{u}(t))^T \left(p(t) + \int_{[0,t)} q_x(s, x^{\bar{u}}(s))^T \nu(ds) \right),$$

$$-\left(p(1) + \int_{[0,1]} q_x(s, x^{\bar{u}}(s))^T \nu(ds) \right) = \phi_x(x^{\bar{u}}(1)) +$$

$$\sum_{i \in E} \alpha_i^1 (h_i^1)_x(x^{\bar{u}}(1)) + \sum_{j \in I} \alpha_j^2 (h_j^2)_x(x^{\bar{u}}(1)),$$

$$\left(p(t)^T + \int_{[0,t)} q_x(s, x^{\bar{u}}(s)) \nu(ds) \right) f_u(t, x^{\bar{u}}(t), \bar{u}(t)) u \leq$$

$$\left(p(t)^T + \int_{[0,t)} q_x(s, x^{\bar{u}}(s)) \nu(ds) \right) f_u(t, x^{\bar{u}}(t), \bar{u}(t)) \bar{u}(t)$$

$$\forall u \in \Omega, \text{ a.e. on } T, \tag{1.15}$$

$$\text{supp}\{\nu\} \subset R_{0,u} \text{ and } \alpha_j^2 = 0 \text{ if } h_j^2(x^{\bar{u}}(1)) < 0. \tag{1.16}$$

Theorem 1.1 *Assume that the data for* (**P**) *satisfies hypotheses* (**H1**), (**H2**), (**H3**) *and* (**CQ**). *Let* $\{u_k\}$ *be a sequence of control functions generated by* Algorithm 2 *and let* $\{c_k\}$ *be the corresponding sequence of penalty parameters. Then*

(i) $\{c_k\}$ *is a bounded sequence,*

(ii)

$$\lim_{k\to\infty} \sigma_{c_k,R_{\epsilon,u_k}}(u_k) = 0, \quad \lim_{k\to\infty} \max_{i\in E} \left|\tilde{h}_i^1(u_k)\right| = 0,$$

(iii) any $\mathcal{L}^2$ accumulation point of $\{u_k\}$ is feasible for (**P**) *and satisfies extremality conditions* (**NC**).

Proof.

From the approximation of R_{ϵ,u_k}, after a finite number of iterations, we have $c_k = c = const$ (c.f. the proof of *Theorem 3.5.1*) and

$$\begin{aligned} \tilde{F}_c(u_{k+1}) - \tilde{F}_c(u_k) &\leq \gamma\alpha_k\sigma_{c,A_k}(u_k), \\ \alpha_k &> \tau_k. \end{aligned} \tag{1.17}$$

As in the proof of *Theorem 3.5.1* we can show that

$$\lim_{k\to\infty} \alpha_k\sigma_{c,A_k}(u_k) = 0. \tag{1.18}$$

Assume $\sigma_{c,A_k}(u_k) \not\to_{k\to\infty} 0$, then $\lim_{k\to\infty}\alpha_k = 0$ but this together with (1.17) implies that for $u_k(\alpha_k/\eta) := P_{\mathcal{U}}[u_k + (\alpha_k/\eta)]$:

$$\tilde{F}_c(u_k(\alpha_k/\eta)) - \tilde{F}_c(u_k) > \gamma(\alpha_k/\eta)\sigma_{c,A_k}(u_k), \quad \text{or} \tag{1.19}$$

$$\max_{t\in T} \tilde{q}(u_k(\alpha_k/\eta))(t) > 0, \quad \text{or} \tag{1.20}$$

$$\max_{j\in I} \tilde{h}_j^2(u_k(\alpha_k/\eta)) > 0. \tag{1.21}$$

Consider the case (1.19). By *Proposition 1.2*

$$\begin{aligned} \tilde{F}_c(u_k(\alpha_k/\eta)) - \tilde{F}_c(u_k) &\leq (\alpha_k/\eta)\sigma_{c,A_k}(u_k) + (\alpha_k/\eta)K\tau_k \\ &\quad +o(\alpha_k), \; K < \infty, \\ \lim_{k\to\infty} \alpha_k^{-1}o(\alpha_k) &= 0, \; \lim_{k\to\infty} \tau_k = 0. \end{aligned} \tag{1.22}$$

Arguing as in the proof of *Theorem 3.5.1*, we arrive at the contradiction

$$0 \leq (1-\gamma)\tilde{\sigma}, \; \tilde{\sigma} < 0, \; \gamma \in (0,1). \tag{1.23}$$

Hence,

$$\lim_{k\to\infty} \sigma_{c,A_k}(u_k) = 0. \tag{1.24}$$

This implies that

$$\lim_{k\to\infty} \sigma_{c,R_{\epsilon,u_k}}(u_k) = 0 \tag{1.25}$$

since

$$A_k \subseteq R_{\epsilon,u_k} \text{ and } \sigma_{c,A_k}(u_k) \leq \sigma_{c,R_{\epsilon,u_k}}(u_k) \leq 0. \tag{1.26}$$

The cases (1.20) and (1.21) can be treated as in the proof of *Theorem 3.5.1* if we notice that $\tau_k \to 0$. The rest of the proof is exactly the same as the relevant part of the proof of *Theorem 3.5.1*, taking as a starting point relationship (1.25). ■

The scheme for the discretization of the state constraint mimics the approximation of a subdifferential of the Lipschitzian function by a finite number of subgradients with the exception that knowledge of the subdifferential of the Chebyshev functional is exploited to much extent (c.f. [64]).

1.1 Second Order Correction To the Line Search

A feature of *Algorithm 2* is that, if we assume that at the kth iteration the control function u_k satisfies the equality constraints, then a new control is drawn from the projection of the ray $\{u_k + \alpha d_k^j : \alpha > 0\}$ on the control constraint set. We can expect that, along the ray, the equality constraint is satisfied only to first order. Improvements to the algorithm's performance can be achieved by replacing the ray by a curve $\{u_k + \alpha d_k^j + \alpha^2 \tilde{d}_k^j : \alpha > 0\}$ along which the (nonlinear) equality constraints are satisfied more accurately. A suitable choice of $\tilde{d}_k^j$ is suggested in [74] in a search procedure whose purpose was to guarantee superlinear convergence of second order algorithms for problems with nonlinear constraints (it was also used in several SQP feasible directions algorithms: [83], [68]). Here $\tilde{d}_k^j$, giving the 'second order correction (SOC) term' $\alpha^2 \tilde{d}_k^j$, is computed from u_k and d_k^j by solving the optimization problem:

$$\min_{d \in \mathcal{U}_{\omega, u_k}} \quad \|d\|^2_{\mathcal{L}^2}$$

subject to

$$\begin{aligned}
\tilde{h}_i^1(u_k + d_k^j) + \left\langle \nabla \tilde{h}_i^1(u_k), d \right\rangle &= 0 \ \forall i \in E \\
\tilde{h}_l^2(u_k + d_k^j) + \left\langle \nabla \tilde{h}_l^2(u_k), d \right\rangle &\leq \sigma_{c_k^j, A_k^j}(u_k) \ \forall l \in I_{\varepsilon, u_k} \\
\tilde{q}(u_k + d_k^j)(t) + \langle \nabla \tilde{q}(u_k)(t), d \rangle &\leq \sigma_{c_k^j, A_k^j}(u_k) \ \forall t \in A_k^j,
\end{aligned}$$

where I_{ε, u_k} indicates the ε–active endpoint inequality constraints at u_k.

$\tilde{d}_k^j$ generated in this way corresponds to a Newton step with respect to the equality constraints at the point $u_k + d_k^j$.[1] If there are no feasible elements for this optimization problem then the second order correction is not applied (i.e. we set $\tilde{d}_k^j = 0$).

---

[1]It was shown in [12] that using $\langle \nabla \tilde{h}_i^1(u_k), d \rangle$, $\langle \nabla \tilde{h}_j^2(u_k), d \rangle$, $\langle \nabla \tilde{q}(u_k)(t), d \rangle$ instead of $\langle \nabla \tilde{h}_i^1(u_k + d_k^j), d \rangle$, $\langle \nabla \tilde{h}_j^2(u_k + d_k^j), d \rangle$, $\langle \nabla \tilde{q}(u_k + d_k^j)(t), d \rangle$ is sufficient.

Algorithm 2 is now modified by replacing $\tilde{u}_k = P_{\mathcal{U}}[u_k + \alpha_k^j d_k^j]$ by $\tilde{u}_k = P_{\mathcal{U}}[u_k + \alpha_k^j d_k^j + (\alpha_k^j)^2 \tilde{d}_k^j]$ in Step 4 of the algorithm. We have observed that the performance of the algorithm is significantly improved by this change. The convergence analysis is uneffected because only second order changes are introduced.

1.2 Resetting the Penalty Parameter

In order to avoid unnecessarily large values of the penalty parameter c_k we should reset it as iterations proceed. The extreme case is to reset the penalty parameter to c_0 at the beginning of each iteration, i.e., in Step 3 of *Algorithm 2* to choose c_k from $\{c^0, \kappa c^0, \kappa^2 c^0, \ldots\}$.

Theorem 1.2 *Assume hypotheses* **(H1)**, **(H2)**, **(H3)** *and* **(CQ)** *are satisfied. Let* $\{u_k\}$ *be a sequence generated by* Algorithm 2, *modified to incorporate resetting of the penalty parameter at every iteration. Assume that* $\{u_k\}$ *is* $\mathcal{L}^2$ *convergent sequence with limit* $\bar{u}$. *Then* $\bar{u}$ *satisfies the condition* **(NC)**.

Proof.

The difference with the earlier analysis is that the penalty parameter c_k can now decrease and increase. However arguments used in the proof of *Theorem 3.5.1* can be used to show $\{c_k\}$ is a bounded sequence in the present case. They can also be used to show that $\bar{u}$ satisfies **(NC)** provided we establish

$$\alpha_k \sigma_{c_k, A_k}(u_k) \to 0 \quad \text{as} \quad k \to \infty. \tag{1.27}$$

(the arguments justifying the related property (1.18) break down when the penalty parameter is not eventually constant.) However, in the present context we can write

$$\tilde{F}_{c_k}(u_{k+1}) - \tilde{F}_{c_k}(u_k) \le \gamma \alpha_k \sigma_{c_k, A_k}(u_k) \le 0 \;\forall k.$$

Since it is hypothesized that the sequence $\{u_k\}$ is $\mathcal{L}^2$ convergent, and $\{c_k\}$ is bounded, it follows that the left side has limit zero as $k \to 0$. (1.27) follows. ■

2 Semi–Infinite Programming Problem

Any computer implementation of the algorithms proposed here requires control and state discretizations. Uniform piecewise constant discretization of control variables was applied with N nodes.

Consider

$$u^N(t) := \sum_{j=1}^{N} \phi_j(t)u^{N,j}, \; t \in T, \; u^{N,j} \in \Omega,$$
$$\phi_j(t) = \begin{cases} 1, & \text{if } t \in [t_{j-1}, t_j] \\ 0, & \text{if } t \notin [t_{j-1}, t_j] \end{cases} \tag{2.1}$$

and $t_j = j/N$, $j \in \{0, 1, \ldots, N\}$.

The functions ϕ_j are linearly independent and we can conclude that the space

$$\mathcal{L}_m^N[T] := \left\{ u \in \mathcal{L}_m^2[T] : \; u(t) = \sum_{j=1}^{N} \phi_j(t)u^{N,j}, \; u^{N,j} \in \mathcal{R}^m \right\}$$

is in one–to–one correspondence with the Euclidean space $\mathcal{R}^{mN}$, so that for any $u^N \in \mathcal{L}_m^N[T]$ there exists the linear, invertible transformation $W_N : \mathcal{L}_m^N[T] \to \mathcal{R}^{mN}$ such that $W_N(\sum_{j=1}^N \phi_j(\cdot)u^{N,j}) := \vec{u}^N := (u^{N,1}, u^{N,2}, \ldots, u^{N,N})$. This means that *Algorithm 2* when applied to the problem **(P)** with $u \in \mathcal{L}_m^N[T] \bigcap \mathcal{U}$ becomes a semi–infinite programming algorithm. In order to verify this we have to derive formulas for $\langle \nabla \tilde{F}_0(u^N), d^N \rangle$, $d^N \in \mathcal{L}_m^N[T]$, etc, to show that they correspond to scalar products in the space $\mathcal{R}^{mN}$.

Let us introduce the adjoint equations for the functional $\tilde{F}_0(u^N)$ defined by $\phi(x^{u^N}(1))$ and ordinary differential equations (3.1.1) (c.f. §2.2):

$$\begin{aligned} p(1) &= \phi_x(x^{u^N}(1))^T \\ \dot{p}(t) &= -f_x(t, x^{u^N}(t), u^N(t))^T p(t), \; \text{a.e. on } T. \end{aligned} \tag{2.2}$$

Next, we define the gradient

$$\nabla \tilde{F}_0(u^N)(t) := f_u(t, x^{u^N}(t), u^N(t))^T p(t).$$

If $d^N \in \mathcal{L}_m^N[T]$, then we can show that

$$\left\langle \nabla \tilde{F}_0(u^N), d^N \right\rangle = \langle \nabla \bar{F}_0(\vec{u}^N), W_N(d^N) \rangle_{\mathcal{R}^{mN}},$$

where

$$\begin{aligned} \nabla \bar{F}_0(\vec{u}^N) &:= (\nabla_1 \bar{F}_0(\vec{u}^N), \ldots, \nabla_N \bar{F}_0(\vec{u}^N)), \\ \nabla_j \bar{F}_0(\vec{u}^N) &:= \int_{t_{j-1}}^{t_j} \nabla \tilde{F}_0(u^N)(t)dt, \; j = 1, \ldots, N. \end{aligned}$$

These formulas should be used in subproblems $\mathbf{P_{c,A}(u)}$ instead of the corresponding $\left\langle \nabla \tilde{F}_0(u), d \right\rangle$, $\left\langle \nabla \tilde{h}_i^1(u), d \right\rangle$, etc. The term $||d||^2_{\mathcal{L}^2}$ should be replaced by $1/N ||W_N(d^N)||^2$.

After the discretization of control functions the problem (**P**) becomes

$$\min_{\vec{u}^N \in \vec{\mathcal{U}}^N} \left\{ \bar{F}_0(\vec{u}^N) := \tilde{F}_0(W_N^{-1}(\vec{u}^N)) \right\}$$

subject to

$$\begin{aligned} \bar{h}_i^1(\vec{u}^N) &:= \tilde{h}_i^1(W_N^{-1}(\vec{u}^N)) = 0 \ \ \forall i \in E \\ \bar{h}_j^2(\vec{u}^N) &:= \tilde{h}_j^2(W_N^{-1}(\vec{u}^N)) \leq 0 \ \ \forall j \in I \\ \bar{q}(\vec{u}^N)(t) &:= \tilde{q}(W_N^{-1}(\vec{u}^N))(t) \leq 0 \ \ \forall t \in T. \end{aligned}$$

Here, W_N^{-1} is the inverse operator to W_N and

$$\begin{aligned} \vec{\mathcal{U}}^N := \{\vec{u}^N = (u^{N,1}, \ldots, u^{N,N}) \in \mathcal{R}^{mN} :\ b_-^i \leq (u^{N,j})_i \leq b_+^i, \\ i \in \{1, \ldots, m\},\ j \in \{1, \ldots, N\}\}. \end{aligned} \tag{2.3}$$

We call this problem ($\mathbf{P^N}$).

Algorithm 2 modified for the problem ($\mathbf{P^N}$) generates a sequence of controls $\{\vec{u}_k^N\}$ and the corresponding sequence of penalty parameters $\{c_k\}$ such that $\{c_k\}$ is bounded (because the set $\vec{\mathcal{U}}^N$ is compact in $\mathcal{R}^{mN}$) and any accumulation point of $\{u_k^N\}$ satisfies optimality conditions (**NC**) adapted for the problem ($\mathbf{P^N}$). This is true under hypotheses (**H1**)–(**H3**) and (**CQ**) also adapted for the problem ($\mathbf{P^N}$).

The semi–infinite programming problem is further discussed in Chapter 4.

3 Numerical Examples

In this section we describe the results of applying *Algorithm 2* to several optimal control problems. All computations were performed on a Sun SPARCstation 10 Model 51. The program was implemented in FORTRAN using BLAS and *double precision* accuracy.

The implementation of *Algorithm 2* requires also the discretization of state constraints. In the first three examples provided below the state constraints were imposed at NX points uniformly distributed over $[0, 1]$. (The discretization of state constraints in the fourth example is discussed later.) If $R_{\varepsilon,u}^d$ represents the discrete representation of $R_{\varepsilon,u}$ due to this discretization of state constraints, then A_k^j's in Step 3 of *Algorithm 2* were based on $(R_{\varepsilon,u_k^N}^d, \xi_k^j)$–uniform approximations (these approximations are further discussed in Chapter 6).

The choice of ε is very important since the number of gradients evaluated by *Algorithm 2* strongly depends on it. *Algorithm 2* is globally convergent for any positive ε; however both large and small values of ε could result either in many gradient evaluations, or in many iterations. We found that the following choice of ε, used in the calculations reported below, is a good compromise:

$$\varepsilon = \varepsilon(\vec{u}_k^N) = \max\{\varepsilon_{min}, 0.1(q_{max}(\vec{u}_k^N) - q_{min}(\vec{u}_k^N))\},$$

where

$$\varepsilon_{min} = 10^{-3}, \quad q_{max}(\vec{u}_k^N) = \max_{t \in T} \bar{q}(\vec{u}_k^N)(t), \quad q_{min}(\vec{u}_k^N) = \min_{t \in T} \bar{q}(\vec{u}_k^N)(t).$$

Notice that we used ε which was adaptively adjusted. Since, for all k, $0 < \varepsilon_{min} \leq \varepsilon(\vec{u}_k^N) < \infty$ it was legitimate.

We also found that if ξ is relatively large then decreasing it in Step 4 of *Algorithm 2* can be largely avoided if A_k^j consists of an $(R^d_{\varepsilon,u_k^N}, \xi_k)$–uniform approximation enlarged by A_{k-1}. We use this approximation of state constraints with $\xi^0 = 0.1$, $\tau^0 = 10^{-4}$. The other parameters used in computations were: $c^0 = 1$, $\eta = 0.5$, $\gamma = 0.003$.

We employed a range-space active set method to solve the direction finding subproblems ([95]). We found that the active set method performs much better than the proximity algorithm described in [75] (even with the modifications due to Hauser ([59])).[2] One reason for good performance of the active set method is the fact that only ω–active box constraints (1.2) are included in the problem $\mathbf{P}_{\mathbf{c},\mathbf{A}}(\mathbf{u})$. In all numerical examples constraints were satisfied with tolerance 10^{-5}. The system and adjoint equations for the first three examples were integrated by the LSODE code (fifth–order BDF method —([60]) with RTOL, ATOL equal to 10^{-6}. The gradients $\nabla \bar{F}_0(\vec{u}^N)$, $\nabla \bar{q}(\vec{u}^N)$, etc, were calculated with the help of adjoint equations which were also integrated by the LSODE code. The penalty parameter was reset every 5 iterations.

Example 1. Our first problem is a minimum time brachistochrone problem with a state variable inequality constraint as well as with a state terminal equality condition. This problem was previously investigated by Polak, Yang and Mayne ([86]). As it was shown in [11] the brachistochrone problem can be stated as follows.

$$\min_{u,t_f} t_f$$

subject to the constraints

$$\dot{x}_1(t) = \sqrt{2gx_2(t)}\cos(u(t))$$

[2]Note that proximity algorithms described in [123] and [75] were proposed for convex problems with an infinite number of constraints. It is well–known that the number of constraints can have very strong effect on the performance of a proximity algorithm. If the number of constraints is finite then modifications proposed by Hauser ([59]) should be applied to increase efficiency of these methods. As far as we know these modifications cannot be applied to proximity methods applied to problems with an infinite number of constraints.

$$\begin{aligned}\dot{x}_2(t) &= \sqrt{2gx_2(t)}\sin(u(t))\\ x_2(t) - x_1(t)\tan(\theta) - h &\le 0,\ t \in [0, t_f],\\ x_1(t_f) &= l,\end{aligned}$$

where g is due to gravity, $\theta = 0.2$, $h = 0.6$, $l = 4.0$. We used following initial conditions: $x_1(0) = 0.0$, $x_2(0) = 0.3$, $(t_f)_0 = 2.0$, $u_0 \equiv 0.0$.

The brachistochrone problem can be reformulated by applying the time transformation:

$$[0, t_f] \ni t \to \tau \in [0, 1] : \ \tau = t/t_f$$

to the problem with the parameter t_f. After the transformation the system equations are as follows

$$\begin{aligned}\dot{x}_1(t) &= t_f\sqrt{2gx_2(t)}\cos(u(t))\\ \dot{x}_2(t) &= t_f\sqrt{2gx_2(t)}\sin(u(t)),\ t \in T.\end{aligned}$$

If an optimal control problem has parameters w as decision variables then gradients of the problem functionals, with respect to the parameters, can be calculated using the adjoint variables needed also for the evaluation of gradients with respect to controls.

Consider the functional $\bar{F}_0(u^N, w) = \phi(x^{u^N,w}(1), w)$—$x$ is now determined by (u^N, w)—and the system equations

$$\begin{aligned}\dot{x}(t) &= f(t, x(t), u^N(t), w),\ x(0) = x_0,\ \text{a.e. on } T. \qquad (3.1)\\ x(0) &= x_0\end{aligned}$$

Then, we have

$$\begin{aligned}\nabla_w \bar{F}_0(u^N, w) &:= \int_T f_w(t, x^{u^N,w}(t), u^N(t), w)^T p(t)dt + \\ &\quad \phi_w(x^{u^N,w}(1), w)^T\end{aligned}$$

where p is the solution to the adjoint equations

$$\begin{aligned}p(1) &= \phi_x(x^{u^N,w}(1), w)^T\\ \dot{p}(t) &= -f_x(t, x^{u^N,w}(t), u^N(t), w)^T p(t),\ \text{a.e. on } T.\end{aligned}$$

The results of computations for the brachistochrone problem are summarized in Table 4.1. The minimum time determined in [86] was equal to 0.99971, however the violation of the endpoint constraint was ≥ 0.03. FUN is the value of the objective function obtained at the last iteration, ITR is the number of iterations, IFUN — the number of function evaluations, IGRD — the number of gradient evaluations, $\sigma_{c,A}(u_{ITR})$ is the last value of the directional derivative, CPU is measured in seconds. In the second

	FUN	ITR, IFUN, IGRD	$\sigma_{c,A}(u_{ITR})$	CPU
N=100 NX=300	1.0016	26, 123, 538	$-1.8 \cdot 10^{-4}$	28.6
N=1000 NX=3000	1.0016	26, 115, 581	$-2.3 \cdot 10^{-4}$	375.0

TABLE 4.1. Example 1, summary of results.

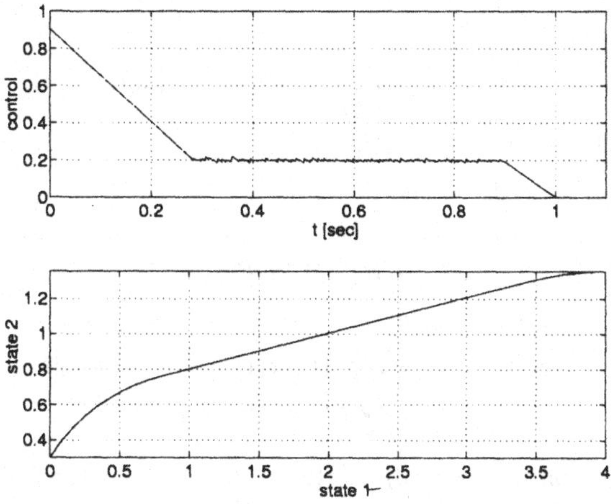

FIGURE 4.2. Example 1, optimal control and state trajectories for $N = 1000$.

experiment, $NX = 3000$, at most 30 points were used to construct A_k for the discretization of R_{ε,u_k}, even though the state constraint was active at 2000 node points associated with the discretization of the state constraint (i.e. on two thirds of the underlying time interval). The history of the computation (including CPU time) of the optimal control for this example, as for the next two, illustrates that the computational time does not 'explode' w.r.t. the dimension of discretization, a feature of *Algorithm 2*, which has been observed on other examples we have tested and which makes it suitable for applications involving many state variables when computation of the relevant gradients is expensive (c.f. problems discussed in Chapter 6).

The calculated control and state trajectories (for the case $N = 1000$) are shown in Figure 4.2. Notice jittery behaviour of the control on an *active arc* due to the fact that *Algorithm 2* performed a finite number of iterations.

	FUN	ITR, IFUN, IGRD	$\sigma_{c,A}(u_{ITR})$	CPU
N=100 NX=400	$5.6888 \cdot 10^{-3}$	40, 119, 855	$-8.42 \cdot 10^{-7}$	125.2
N=1000 NX=4000	$5.6798 \cdot 10^{-3}$	41, 125, 1023	$-8.99 \cdot 10^{-7}$	1398.0

TABLE 4.2. Example 2, summary of results.

Example 2. This is the crane problem earlier studied in [118].

$$\begin{aligned}
\min_u \quad & 0.5 \int_0^{t_f} \left\{ [x_3(t)]^2 + [x_6(t)]^2 \right\} dt \\
\text{s. t.} \quad \dot{x}_1(t) &= x_4(t) \\
\dot{x}_2(t) &= x_5(t) \\
\dot{x}_3(t) &= x_6(t) \\
\dot{x}_4(t) &= u_1(t) + 17.625 x_3(t) \\
\dot{x}_5(t) &= u_2(t) \\
\dot{x}_6(t) &= -1/x_2(t) \left[u_1(t) + 27.0756 x_3(t) + \right. \\
& \left. 2.0 x_5(t) x_6(t) \right], \ t \in [0, t_f].
\end{aligned}$$

Take endpoint conditions $x(0) = (0.0, 22., 0.0, 0.0, -0.8, 0.0)^T$, $x(t_f) = (10, 14, 0.0, 2.5, 0.0, 0.0)^T$, state constraints

$$\begin{aligned}
|x_4(t)| &\leq 2.5, \quad t \in [0, t_f], \\
|x_5(t)| &\leq 1.0, \quad t \in [0, t_f]
\end{aligned}$$

and control constraints

$$\begin{aligned}
|u_1(t)| &\leq 2.83374, \quad t \in [0, t_f], \\
-0.80865 \leq u_2(t) &\leq 0.71265, \quad t \in [0, t_f].
\end{aligned}$$

For $t_f = 9.0$ and the initial controls $(u_0)_1(\cdot) \equiv 0.0$, $(u_0)_2(\cdot) \equiv 0.0$ we obtained results summarized in Table 4.2. The sets A_k consisted of at most 8 points for each state constraint in the problem. The initial state for the problem is different from that used in [118]: $x_5(0) = -0.8$ instead of $x_5(0) = -1.0$ since the initial conditions stated in [118] do not guarantee that the constraint qualification (**CQ**) is satisfied.

Example 3 The following problem was initially considered by Logsdon

	FUN	ITR, IFUN, IGRD	$\sigma_{c,A}(u^{ITR})$	CPU
N=100 NX=400	1.0097	29, 31, 150	0.0	22.4
N=1000 NX=4000	0.96809	29, 31, 150	$-7.74 \cdot 10^{-7}$	227.2

TABLE 4.3. Example 3, summary of results.

([69]) and then by Wright ([126],[127]):

$$\min_u \sum_{i=1}^{4} [x_i(t_f)]^2 \tag{3.2}$$

subject to the constraints:

$$\begin{aligned}
\dot{x}_1(t) &= -0.5x_1(t) + 5x_2(t) \\
\dot{x}_2(t) &= -5x_1(t) - 0.5x_2(t) + u(t) \\
\dot{x}_3(t) &= -0.6x_3(t) + 10x_4(t) \\
\dot{x}_4(t) &= -10x_3(t) - 0.6x_4(t) + u(t) \\
x_i(t_f) &\leq 1, \ i = 1, \ldots, 4 \\
|u(t)| &\leq 1, \ t \in [0, t_f],
\end{aligned}$$

where $t_f = 4.2$. For the initial states: $x_i(0) = 10$, $i = 1, \ldots, 4$, and the initial control: $u_0 \equiv 0$ we found the approximations to the solution summarized in Table 4.3. The final objective value found in [69] was 1.00347, while that in [126] was 1.0126 (1.00357 in [127] when post optimization integration with the final u was used to obtain the final function value). The solution is of *bang-bang* type on the whole horizon and is shown, for $N = 1000$, in Figure 4.3. The switching points, for $N = 100$, were located at 0.126, 0.882, 1.386, 2.184, 2.604, 3.444, 3.822. For $N = 1000$ they were: 0.1092, 0.8904, 1.365, 2.163, 2.6208, 3.4356, 3.8766. The discrepency (quite significant) in switching points explains much better value of the performance index for $N = 1000$.

It is interesting to compare computing times recorded by three programs tested on this example. The algorithm of Logsdon required over 6 hours of CPU time on a VAX 6320.[3] The method by Wright needed 73.7 seconds of CPU time on a Solbourne 5E/900 with 256 Mbytes of main mem-

[3]The excessive computing time can be explained by the fact that gradients

ory ([126]).[4] The use of the projection $P_{\mathcal{U}}[\cdot]$ in *Algorithm 2* and a primal range–space active set method to solve $\mathbf{P_{c,A}(u)}$ (discussed in the Appendix) helped to reduce the computing time in our approach.

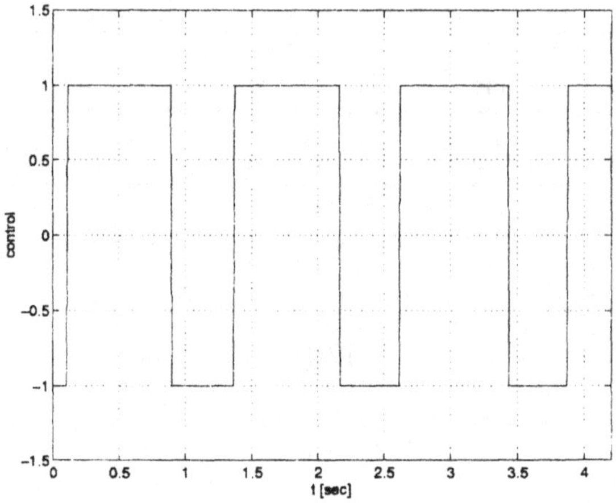

FIGURE 4.3. Example 2, optimal control for $N = 1000$.

Example 4. The fourth problem was analysed in [77], [27] and [28]. The problem is related to the landing of a passenger aircraft in the presence of windshear.

The system dynamics is described by the following equations:

$$\begin{aligned}
\dot{x} &= V\cos\gamma + W_x && (3.3)\\
\dot{h} &= V\sin\gamma + W_h && (3.4)\\
\dot{V} &= (T/m)\cos(\alpha+\delta) - \\
&\quad D/m - g\sin\gamma - (\dot{W}_x\cos\gamma + \dot{W}_h\sin\gamma) && (3.5)\\
\dot{\gamma} &= (T/mV)\sin(\alpha+\delta) + L/mV - \\
&\quad (1/V)g\cos\gamma + (1/V)(\dot{W}_x\sin\gamma - \dot{W}_h\cos\gamma) && (3.6)\\
\dot{\alpha} &= u,\ t\in[0,t_f] && (3.7)
\end{aligned}$$

where x is the horizontal distance, h the altitude, V the relative velocity and γ the relative path inclination. α, the relative angle of the attack (which is the proper control), is also regarded as a state variable which has to satisfy

of the problem were evaluated with the help of sensitivity equations (briefly discussed in Chapter 6). Obviously the problem requires the approximation of controls by piecewise constant functions with the parameter N large enough to locate accurately switching times. This results in a large number of sensitivity equations.

[4]Wright needed 20.3 seconds of CPU time on a Cray Y-MP to find value of the objective function 1.00357 using some version of his algorithm—[127].

the constraints

$$
\begin{aligned}
-\dot{\alpha}_{max} &\leq \dot{\alpha}(t) \leq \dot{\alpha}_{max} && (3.8)\\
0 &\leq \alpha(t) \leq \alpha_{max}, \quad t \in [0, t_f] && (3.9)
\end{aligned}
$$

and the first constraint can be directly imposed on the new control variable u.

The approximation of the aerodynamics forces acting on the aircraft: T —thrust, D—drag and L—lift are represented by the equations

$$
\begin{aligned}
T &= \beta T_\star && (3.10)\\
T_\star &= A_0 + A_1 V + A_2 V^2 && (3.11)\\
D &= (1/2) C_D(\alpha)\rho S V^2 && (3.12)\\
C_D(\alpha) &= B_0 + B_1\alpha + B_2\alpha^2 && (3.13)\\
L &= (1/2) C_L(\alpha)\rho S V^2 && (3.14)\\
C_L(\alpha) &= \begin{cases} C_0 + C_1\alpha & \text{if } 0 \leq \alpha \leq \alpha_\star \\ C_0 + C_1\alpha + C_2(\alpha - \alpha_\star)^2 & \text{if } \alpha_\star \leq \alpha \leq \alpha_{max} \end{cases}
\end{aligned}
$$

The power setting β (another proper control) is specified in advance as in [77] and [27]:

$$
\beta(t) = \begin{cases} \beta_0 + \dot{\beta} t & \text{if } 0 \leq t \leq t_0 \\ 1 & \text{if } t_0 \leq t \leq t_f. \end{cases}
$$

The windshear model is given by the following velocity components:

$$
\begin{aligned}
W_x &= kA(x) && (3.15)\\
W_h &= k(h/h_\star)B(x), && (3.16)
\end{aligned}
$$

where the functions $A(x)$ and $B(x)$ are as described in [27]:

$$
A(x) = \begin{cases} -50 + ax^3 + bx^4 & \text{if } 0 \leq x \leq 500 \\ (1/40)(x - 2300) & \text{if } 500 \leq x \leq 4100 \\ 50 - a(4600 - x)^3 - b(4600 - x)^4 & \text{if } 4100 \leq x \leq 4600 \\ 50 & \text{if } 4600 \leq x, \end{cases}
$$

$$
B(x) = \begin{cases} dx^3 + ex^4 & \text{if } 0 \leq x \leq 500 \\ -51\exp[-c(x - 2300)^4] & \text{if } 500 \leq x \leq 4100 \\ d(4600 - x)^3 + e(4600 - x)^4 & \text{if } 4100 \leq x \leq 4600. \end{cases}
$$

Because the functions $A(x)$ and $B(x)$ are not continuously differentiable at three points we used quadratic splines to smooth them around these points. This resulted in a slightly different model to that used in [28] but we believe, and the numerical results presented below support our view,

$t_f = 40$ sec	$\rho = 0.2203\times10^{-2}$ lb sec^{2}ft^{-4}
$h_R = 1000$ ft	$S = 0.1560\times10^{4}$ ft^2
$\dot{\alpha}_{max} = 3$ deg sec^{-1}	$g = 3.2172\times10^{1}$ ft sec^{-2}
$\alpha_{max} = 17.2$ deg	$mg = 150\,000$ lb
$\delta = 2$ deg	$h_\star = 1000$ ft
$\beta_0 = 0.3825$	$\dot{\beta}_0 = 0.2$ sec^{-1}
$t_0 = (1-\beta_0)/\dot{\beta}_0$	$k = 1$
$x_0 = 0$ ft	$\gamma_0 = -2.249$ deg
$h_0 = 600$ ft	$\alpha_0 = 7.353$ deg
$V_0 = 239.7$ ft sec^{-1}	$\gamma_f = 7.431$ deg
$A_0 = 0.4456\times10^{5}$ lb	$A_1 = -0.2398\times10^{2}$ lb sec^{-1}
$A_2 = 0.1442\times10^{-1}$ lb sec^{2}ft^{-2}	$B_0 = 0.1552$
$B_1 = 0.12369$ rad^{-1}	$B_2 = 2.4203$ rad^{-2}
$C_0 = 0.7125$	$C_1 = 6.0877$ rad^{-1}
$C_2 = -9.0277$ rad^{-2}	$\alpha_\star = 12$ deg
$a = 6\times10^{-8}$sec^{-1}ft^{-2}	$b = -4\times10^{-11}$sec^{-1}ft^{-3}
$c = -\ln(25/30.6)\times10^{-12}$ft^{-4}	$d = -8.02881\times10^{-8}$sec^{-1}ft^{-2}
$e = 6.28083\times10^{-11}$sec^{-1}ft^{-3}	

TABLE 4.4. Parameters of the windshear problem.

that these differences are not significant. All values of parameters for the problem were taken from [27]. They are given in Table 4.4.

The objective functionals considered in [77], [27] and [28] can be expressed as

$$J(u)[\Lambda,\Theta] = \Lambda \int_0^{t_f} [h_R(t) - h(t)]^6\, dt + \Theta \max_{0\le t\le t_f} [h_R(t) - h(t)]. \quad (3.17)$$

In addition to the state constraints an endpoint constraint was imposed on the relative path inclination:

$$\gamma(t_f) = \gamma_f. \quad (3.18)$$

Miele *at al* ([77]) used a gradient restoration algorithm to minimize $J(u)[1,0]$. Bulirsh *et al* ([27],[28]) solved a number of problems described by (3.3)–(3.16) and with objective functionals related to (3.17). First, they solved problems with the functional $J(u)[1,0]$ and without state constraints (3.9). They used multiple shooting method to solve multipoint boundary value problems defined by necessary optimality conditions. Eventually they solved the problem with the objective functional $J(u)[0,1]$ and state constraints (3.9). Several problems had to be solved to provide a good initial guess of the control parametrization and of the switching structure associated with a sequence of switching points and their corresponding control laws between every pair of adjacent switching or boundary points.

Our feasible directions algorithm is very suitable for the problem because the model of the aircraft is valid if the drag D and the lift L are defined by α which satisfies constraints (3.9). The problem can be transformed to the Bolza problem by introducing a new state variable, a new state constraint and the parameter π:

$$\begin{aligned} \dot{\xi} &= 0, \ \xi(0) = \pi, \\ h_R - h(t) - \xi(t) &\leq 0, \ \ t \in [0, t_f] \\ J(u) &= \xi(t_f). \end{aligned}$$

The solution reported in [28] is very accurate. In order to obtain a similarly accurate control we used the implicit Runge–Kutta method, Radau IIA, with absolute and relative tolerances set to 10^{-8}. The implementation of the integration procedure is described in Chapter 6 together with more detailed description of an optimization procedure based on it.

The discretization of state variables (and state constraints) was determined by the integration procedure and resulted in 1397 point representation of each state trajectory at the last iteration. Despite this very fine discretization of state constraints we needed only an average number of 47 gradients evaluations for $\mathbf{P}_{\mathbf{c},\mathbf{A}}(\mathbf{u})$ subproblems (44 at the last iteration). This highlights the efficiency of the method.

The qualitative behaviour of the calculated state trajectories is very similar to that observed in [28]. The main difference is a much shorter *singular subarc* in the interval $[0, t_1]$ where t_1 is the entry point for the *active arc* of α. In $[0, t_1]$ there are two downward movements of α (accompanying by two upward movements) instead of one (shorter) as reported in [28] (longer downward movement of α must be counterbalanced by an upward movement). The discrepancy in the profiles of α suggests that the problem, as stated in [27] and [28] is not well–posed to the extent that different control strategies can result in the same value of the objective functional. It was pointed out in [28] that the profile of α after the *touch* point for h was reached, can be arbitrary under the condition that the terminal equality constraint for γ is satisfied. It seems that this remark applies also to the subinterval $[0, t_1]$, something which is not unexpected for problems with the Chebyshev objective functional.

The optimal state trajectories are shown in Figs. 4.4–4.5. The $\sigma_{c,A}(u)$ at the last iteration was $\approx -2.0 \cdot 10^{-7}$ and the terminal condition (3.18) was satisfied with the accuracy $1.5 \cdot 10^{-6}$. None of state constraints was violated. The minimum altitude obtained was equal 502.1908 and occurred at one touch point equal to 26.0250 (the other local minimum for h was equal 502.1916 at the time 14.5380). This result is very close to that obtained in [28]: 502.156278 at 25.9973.[5] The discrepancy can be explained by the slightly different model of windshear and the fact that control functions were substituted by piecewise constant approximations imposed on $N = 100$ subintervals. Sets A to approximate $R_{\varepsilon,u}$ were constructed in similar way as for the first two examples with the exception that the state constraints were imposed at every step of the integration procedure. The initial control was $u \equiv 0.001$ and was feasible w.r.t. the state constraints.

Algorithm 2 performed well on the problem. It required 41 iterations, 119 systems equations integrations and the solution of 1921 adjoint equations.

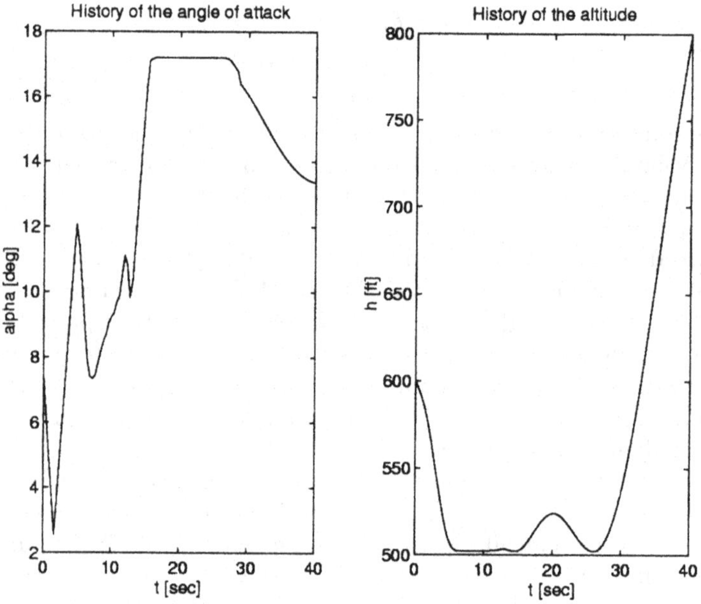

FIGURE 4.4. The windshear problem, optimal state trajectories.

[5]The difference in the altitude is about 1 cm!

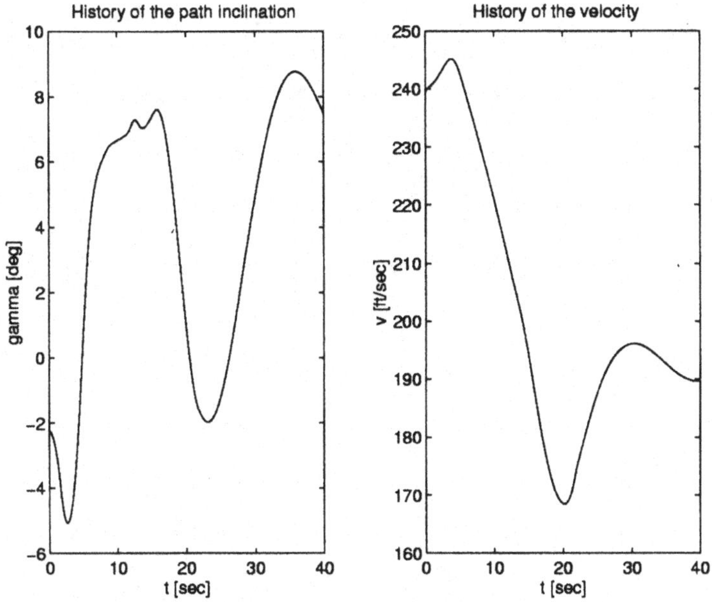

FIGURE 4.5. The windshear problem, optimal state trajectories.

5

Second Order Method

Control functions are substituted by piecewise constant controls. The control problem becomes then a semi-infinite programming problem.

The search direction subproblems have only a finite number of constraints and their Lagrange multipliers are numbers.

1 Introduction

In Chapters 3–4 a feasible directions algorithm for problem (**P**) has been proposed and, in Chapter 4, it has been shown that the algorithm can cope well with problems in which the discretization of control variables is greater than 1000. However, the success of that method very often depends on *a priori* diagonal scaling of control variables (c.f. [107]) and although the scaling, in most cases, is not difficult to guess, this is a drawback of the method. Therefore, we have asked the following question. Is it possible to extend the diagonal scaling to the nondiagonal one, or in other words, is it possible to construct a second order method for control problems with state constraints?

Our approach to a second order method relies on some schemes applied in the first order algorithms presented in Chapters 3–4. Essentially it is an approach in which

(i) problem (3.1.3)–(3.1.6) is solved instead of the problem (**P**),

(ii) search directions use only ε–active state constraints.

These two features of our approach distinguish it from the approach to optimal control algorithms ([2]) which is based on accessory problems. Accessory problems depend on the Lagrange multipliers of state constraints which can have very complicated structure ([27]). Furthermore, in general, accessory problems are nonconvex optimal control problems with state constraints, therefore there are only local convergence results for the approach based on them ([2]).

In this chapter we pay special attention to the problem ($\mathbf{P^N}$) which we recall here:

$$\min_{\vec{u}\in\vec{\mathcal{U}}} \bar{F}_0(\vec{u})$$

subject to the constraints

$$\begin{aligned} \bar{h}_i^1(\vec{u}) &= 0 \;\; \forall i \in E \\ \bar{h}_j^2(\vec{u}) &\leq 0 \;\; \forall j \in I \\ \bar{q}(\vec{u})(t) &\leq 0 \;\; \forall t \in T. \end{aligned}$$

Notice that contrary to the notation used in Chapter 4 we have dropped out superscript from $\vec{u}^N$ and $\vec{\mathcal{U}}^N$. This simplifies notation for the case when N is fixed.

If we assume that $\bar{\vec{u}}$ is a solution to ($\mathbf{P^N}$), $\vec{\mathcal{U}} = \mathcal{R}^{mN}$ and $E(\bar{\vec{u}})$, $I(\bar{\vec{u}})$, $T(\bar{\vec{u}})$ are subsets of E, I and T respectively and such that $\nabla\bar{h}_i^1(\bar{\vec{u}})$, $i \in E(\bar{\vec{u}})$, $\nabla\bar{h}_j^2(\bar{\vec{u}})$, $j \in I(\bar{\vec{u}})$, $\nabla\bar{q}(\bar{\vec{u}})(t)$, $t \in T(\bar{\vec{u}})$ are linearly independent, then we will show that instead of solving the problem ($\mathbf{P^N}$) we can look, locally around $\bar{\vec{u}}$, for a solution of the problem:

$$\min_{\vec{u}} \bar{F}_0(\vec{u})$$

subject to the constraints

$$\begin{aligned} \bar{h}_i^1(\vec{u}) &= 0 \;\; \forall i \in E(\bar{\vec{u}}) \\ \bar{h}_j^2(\vec{u}) &\leq 0 \;\; \forall j \in I(\bar{\vec{u}}) \\ \bar{q}(\vec{u})(t) &\leq 0 \;\; \forall t \in T(\bar{\vec{u}}). \end{aligned}$$

We call this problem ($\mathbf{P_{NLP}(\bar{\vec{u}})}$). Since $\vec{u} \in \mathcal{R}^{mN}$ the number $|E(\bar{\vec{u}})| + |I(\bar{\vec{u}})| + |T(\bar{\vec{u}})|$ is finite.

Precise conditions under which the problem ($\mathbf{P_{NLP}(\bar{\vec{u}})}$) can be solved instead of the problem ($\mathbf{P^N}$) is the subject of this chapter. The notable feature of our method is that it generates a sequence of controls $\{\vec{u}_k\}$ which superlinearly converges to a local solution $\bar{\vec{u}}$. Furthermore, this is guaranteed without assuming that at any local solution the function $\bar{q}(\bar{\vec{u}})(\cdot)$ has only a finite number of maximum points (as it is in other second order methods for semi-infinite programming (SIP) problems: e.g., [52], [130]). In general, second order methods for SIP are constructed under the condition that SIP are reduced to NLP problems at all local solutions. The method presented in this chapter reduces SIP problems to NLP problems at all local solutions. The possibility of a similar reduction in minimax problems is considered in [53] (the paper has been recently drawn to the author's attention by William Hager). In contrast to algorithms analysed in [53] the method introduced in this chapter is globally convergent.

Moreover, our method will generate a superlinearly convergent sequence, if applied to the following 'degenerate' problems:

(i) NLP problems which, at solutions, have active constraints with linearly dependent gradients.

(ii) Finely discretized SIP problems as discussed in [130]. Note that the method presented in [130] is superlinearly convergent under the assumption that the number of active constraints at any solution is not greater than the dimension of the decision vector increased by one (Assumption 4 in [130], p. 470).

(iii) Control problems with state constraints and time independent decision variables (parameters).

For the problem (**P**) we construct a second order algorithm whose direction finding subproblem has a quadratic term incorporating second order terms of the problem functions. This quadratic term is in the form of the Hessian of a certain functional defined by $\tilde{F}_0(u)$, $\tilde{h}_i^1(u)$, $i \in E$, $\tilde{h}_j^2(u)$, $j \in I$, $\tilde{q}(u)(t)$, $t \in T$. Its evaluation is computationally expensive (especially for problems with many state variables—§2.2) therefore the approach taken here is first to transform the problem (**P**) to the problem ($\mathbf{P^N}$) and then to approximate the corresponding Hessian matrix by using a quasi-Newton formula, or any other scheme for approximating the Hessian matrix.

The computational method presented in this chapter is similar to the feasible directions algorithm in the sense that $\tilde{q}(u)(t)$, $t \in T$ are approximated by a finite number of constraints. This is the subject of the next two sections where we analyse the global convergence of our method. The proofs of convergence results stated there are very similar to the relevant proofs presented in Chapters 3–4 and are therefore omitted. These results are needed to introduce the reader to the main section of the chapter, §5.4, where the superlinear rate of convergence of a sequence generated by our algorithm is established.

In order to solve the problem (**P**) we apply an exact penalty function to transform the original problem to a problem with only simple constraints on controls. Therefore, instead of solving the problem (**P**), we look at the problem ($\mathbf{P_c}$):

$$\begin{aligned}\min_{u\in\mathcal{U}}\Big[\tilde{F}_c(u) \quad &:= \quad \tilde{F}_0(u) + c\max\Big[0, \max_{i\in E}\left|\tilde{h}_i^1(u)\right|, \\ &\max_{j\in I}\tilde{h}_j^2(u), \max_{t\in T}\tilde{q}(u)(t)\Big]\Big] \qquad (1.1)\end{aligned}$$

where $c > 0$ is a sufficiently large positive number.

Notice that the exact penalty function is different from that considered in Chapters 3–4. First of all, it is defined by all constraints including inequality constraints. Secondly, it is constructed by multiplying the violation of constraints by a penalty parameter instead of dividing the objective functional by it. The reason for the second change is a more convenient statement of a direction finding subproblem, its quadratic term does not depend then on the penalty parameter.

2 Function Space Algorithm

At each iteration of the exact penalty function algorithm search directions are generated by solving a simplified version of the problem in which the dynamics, cost functional and constraint functionals are replaced by their first order approximations around the current control function u. A quadratic term of the search directions subproblem incorporates second order terms of the functions of the problem (**P**).

For fixed c, H and u the direction finding subproblem for the problem $\mathbf{P_c}$, denoted by $\mathbf{P_c^H(u)}$, is:

$$\min_{d \in \mathcal{U}-u, \beta \in \mathcal{R}} \left[\left\langle \nabla \tilde{F}_0(u), d \right\rangle + c\beta + 1/2 \langle H(d), d \rangle \right]$$

subject to

$$\begin{aligned} \left| \tilde{h}_i^1(u) + \left\langle \nabla \tilde{h}_i^1(u), d \right\rangle \right| &\leq \beta \;\; \forall i \in E \\ \tilde{h}_j^2(u) + \left\langle \nabla \tilde{h}_j^2(u), d \right\rangle &\leq \beta \;\; \forall j \in I \\ \tilde{q}(u)(t) + \langle \nabla \tilde{q}(u)(t), d \rangle &\leq \beta \;\; \forall t \in R_{\varepsilon,u}. \end{aligned}$$

Here, $H : \mathcal{L}_m^2[T] \to \mathcal{L}_m^2[T]$ is a self–adjoint operator which satisfies the following condition.

(BH): There exist numbers $0 < \nu_1 < \nu_2 < \infty$ such that

$$\nu_1 \|d\|_{\mathcal{L}^2}^2 \leq \langle H(d), d \rangle \leq \nu_2 \|d\|_{\mathcal{L}^2}^2 \quad \forall d \in \mathcal{L}_m^2[T]. \tag{2.1}$$

The subproblem can be reformulated as an optimization problem over the space $\mathcal{L}_m^2[T]$ with the objective function which is strictly convex. The problem therefore has the unique solution $(\bar{d}, \bar{\beta})$. Since this solution depends on c, H and u, we may define descent function $\sigma_c^H(u)$ and penalty test function $t_c^H(u)$ (in similar way as in Chapter 3) to be used to test optimality of u and to adjust c, respectively, as

$$\sigma_c^H(u) := \left\langle \nabla \tilde{F}_0(u), \bar{d} \right\rangle + c \left[\bar{\beta} - M(u) \right]$$

and

$$t_c^H(u) := \sigma_c^H(u) + M(u)/c.$$

Here,

$$M(u) := \max \left[0, \max_{i \in E} \left| \tilde{h}_i^1(u) \right|, \max_{j \in I} \tilde{h}_j^2(u), \max_{t \in T} \tilde{q}(u)(t) \right],$$

A function space algorithm is as follows.

Algorithm 3 Fix parameters: $\varepsilon > 0,\ \gamma,\ \eta \in (0,1),\ c^0 > 0,\ \kappa > 1$, operators $H_k,\quad k = 0, 1, \ldots$

1. Choose the initial control $u_0 \in \mathcal{U}$. Set $k = 0$, $c_{-1} = c^0$.
2. Let c_k be the smallest number chosen from $\{c_{k-1}, \kappa c_{k-1}, \kappa^2 c_{k-1}, \ldots\}$ such that the solution (d_k, β_k) to the direction finding subproblem $\mathbf{P}_{\mathbf{c_k}}^{\mathbf{H_k}}(\mathbf{u_k})$ satisfies
$$t_{c_k}^{H_k}(u_k) \leq 0. \tag{2.2}$$
If $\sigma_{c_k}^{H_k}(u_k) = 0$ then STOP.
3. Let α_k be the largest number chosen from the set $\{1, \eta, \eta^2, \ldots, \}$ such that
$$u_{k+1} = u_k + \alpha_k d_k$$
satisfies the relation
$$\tilde{F}_{c_k}(u_{k+1}) - \tilde{F}_{c_k}(u_k) \quad \leq \quad \gamma \alpha_k \sigma_{c_k}^{H_k}(u_k).$$
Increase k by one. Go to Step 2.

The descent function $\sigma_{c_k}^{H_k}(u_k)$ is nonpositive valued at each iteration. This follows from the relation

$$\left\langle \nabla \tilde{F}_0(u_k), d_k \right\rangle + c_k \beta_k + 1/2 \left\langle H(d_k), d_k \right\rangle \leq c_k M(u_k)$$

which holds because $0 \in \mathcal{U} - u_k$. Thus, we have

$$\left\langle \nabla \tilde{F}_0(u_k), d_k \right\rangle + c_k \left[\beta_k - M(u_k)\right] \leq -1/2 \left\langle H(d_k), d_k \right\rangle \leq 0.$$

If some subsequences $\{u_k\}_{k\in K}$, $\{c_k\}_{k\in K}$ of the sequences of control functions and penalty parameters generated by the algorithm have limit points $\bar{u}$ and $\bar{c}$, we would expect that $\sigma_{c_k}^{H_k}(u_k) \to \bar{\sigma}$ and $\bar{\sigma} \geq 0$, a condition which asserts that the direction finding subproblem for $c_k = \bar{c}$ and $u_k = \bar{u}$, has the solution $(d = 0, \beta = 0)$ and which can be interpreted as a first order optimality condition satisfied by $\bar{u}$.

Before concluding the presentation of the algorithm we need to show that Steps 2–3 of *Algorithm 3* can always be carried out. These gaps are filled by the following proposition.

Proposition 2.1 *(i) Assume that hypotheses* (**H1**), (**H2**) *and* (**CQ**) *are satisfied. Then, for any* $u \in \mathcal{U}$, H, *which satisfies* (**BH**), *there exists* $\bar{c} > 0$ *such that for all* $c > \bar{c}$

$$t_c^H(u) \leq 0.$$

(ii) Assume that hypotheses **(H1)** *and* **(H2)** *are satisfied. Then for any* $u \in \mathcal{U}$, H, *which satisfies* **(BH)**, *and* $c > 0$ *for which* $\sigma_c^H(u) < 0$ *there exists* $\bar{\alpha} > 0$ *such that if* $\alpha \in [0, \bar{\alpha})$ *then*

$$\tilde{F}_c(\tilde{u}) - \tilde{F}_c(u) \quad \leq \quad \gamma\alpha\sigma_c^H(u),$$

where $\tilde{u} = u + \alpha d$, $\sigma_c^H(u) = \langle \nabla \tilde{F}_0(u), d \rangle + c[\beta - M(u)]$ *and* (d, β) *is the solution to the direction finding subproblem* $\mathbf{P_c^H(u)}$.

Proof.

The proposition can be proved in the same way as *Proposition 3.4.1.* ∎

The convergence result is as follows.

Theorem 2.1 *Assume that the data for* **(P)** *satisfies hypotheses* **(H1)**, **(H2)**, **(CQ)** *and* H_k *satisfies* **(BH)** *for every* $k = 0, 1, \ldots$. *Let* $\{u_k\}$ *be a sequence of control functions generated by* Algorithm 3 *and let* $\{c_k\}$ *be the corresponding sequence of penalty parameters. Then*

(i) $\{c_k\}$ *is a bounded sequence,*

(ii)

$$\lim_{k \to \infty} \sigma_{c_k}^{H_k}(u_k) = 0, \quad \lim_{k \to \infty} M(u_k) = 0.$$

(iii) Let $\bar{u}$ *be the limit point of any* $\mathcal{L}^2$ *convergent subsequence of* $\{u_k\}$. *Then* $\bar{u}$ *is feasible for* **(P)** *and satisfies extremality conditions* **(NC)** *((4.1.15)–(4.1.16)).*

If we relate this result to well–known results on the convergence of SQP algorithms for problems defined in finite dimensional spaces, we can say that *Algorithm 3* is well–defined. In a finite dimensional space **(CQ)**, needed to prove the assertion *(i)*, is not applicable. However, the set $\vec{\mathcal{U}}$ (see (4.2.3)), which should be considered instead of $\mathcal{U}$, is then compact and this is sufficient to guarantee that $\{c_k\}$ is bounded.

3 Semi–Infinite Programming Method

Algorithm 3 offers a viable alternative to a first order method if the operator H is the Hessian (or its approximation) of the Lagrangian of the subproblem $\mathbf{P_c(u)}$ for u in some neighbourhood of a local solution to the problem **(P)**. In Chapter 2 we have showed that the evaluation of the Hessian of the functional defined by systems equations (3.1.1) requires solving an infinite number of adjoint equations. This effort can be significantly reduced if control functions are substituted by their piecewise constant approximations. This results in the problem $(\mathbf{P^N})$.

The problem ($\mathbf{P^N}$) has been briefly discussed in Chapter 4. Detailed convergence analysis of the implementable version of *Algorithm 3* requires the introduction of a finite approximation of the set R_{ε,u^N}.

Define as in Chapter 4:

$$\begin{aligned} \mathcal{A}^{\xi}_{\varepsilon,\vec{u}} &:= \mathcal{A}^{\xi}_{\varepsilon,u^N} \\ &= \left\{\text{finite sets } A \subset T: \ A \subset R_{\varepsilon,\vec{u}} := R_{\varepsilon,u^N} \subset A + \xi[-1,1]\right\}, \end{aligned}$$

$\xi > 0$.

Denote by $\mathbf{P^H_{c,A}(\vec{u})}$ the modification of the direction finding subproblem $\mathbf{P^H_c(u)}$, adapted for the problem ($\mathbf{P^N}$), in which A replaces $R_{\varepsilon,\vec{u}}$:

$$\min_{\vec{d}\in\vec{\mathcal{U}}-\vec{u},\beta\in\mathcal{R}} \left[\left\langle \nabla \bar{F}_0(\vec{u}), \vec{d}\,\right\rangle + c\beta + 1/2(\vec{d})^T H \vec{d}\,\right]$$

subject to

$$\left|\bar{h}^1_i(\vec{u}) + \left\langle \nabla \bar{h}^1_i(\vec{u}), \vec{d}\,\right\rangle\right| \leq \beta \ \ \forall i \in E \tag{3.1}$$

$$\bar{h}^2_j(\vec{u}) + \left\langle \nabla \bar{h}^2_j(\vec{u}), \vec{d}\,\right\rangle \leq \beta \ \ \forall j \in I \tag{3.2}$$

$$\bar{q}(\vec{u})(t) + \left\langle \nabla \bar{q}(\vec{u})(t), \vec{d}\,\right\rangle \leq \beta \ \ \forall t \in A. \tag{3.3}$$

where, in order to simplify the future notations, $\langle\cdot,\cdot\rangle$ is understood to be $\langle\cdot,\cdot\rangle_{\mathcal{R}^{mN}}$. Here, H is a symmetric matrix which satisfies the following condition.

($\mathbf{BH^N}$): There exist numbers $0 < \nu_1 < \nu_2 < \infty$ such that

$$\nu_1 \left\|\vec{d}\,\right\|^2 \leq (\vec{d})^T H \vec{d} \leq \nu_2 \left\|\vec{d}\,\right\|^2 \quad \forall \vec{d} \in \mathcal{R}^{mN}.$$

Furthermore, we define

$$\begin{aligned} \sigma^H_{c.A}(u) &:= \left\langle \bar{F}_0(\vec{u}), \vec{d}\,\right\rangle + c\left[\beta - M(\vec{u})\right], \\ t^H_{c,A}(\vec{u}) &:= \sigma^H_{c,A}(\vec{u}) + M(\vec{u})/c, \end{aligned}$$

where $(\vec{d},\beta)$ is the solution to the subproblem $\mathbf{P^H_{c,A}(\vec{u})}$.

Eventually, we can state our algorithm for the problem ($\mathbf{P^N}$).

Algorithm 4 Fix parameters: $\varepsilon > 0$, γ, $\eta \in (0,1)$, $0 < c^0 < \infty$, $\kappa > 1$, $\tau^0 > 0$, $\xi^0 > 0$, $\omega > 1$, H_k, $k = 0, 1, \ldots$

1. Choose the initial control $\vec{u}_0 \in \vec{\mathcal{U}}$. Set $k = 0$, $c_{-1} = c^0$, $\xi_0 = \xi^0$, $\tau_0 = \tau^0$.

2. Set $j = 1$. Let $\tau^j_k = \tau_k$, $\xi^j_k = \xi_k$.

3. Choose $A_k^j \in \mathcal{A}_{\varepsilon,\vec{u}_k}^{\xi_k^j}$. Let c_k^j be the smallest number chosen from $\{c_{k-1}, \kappa c_{k-1}, \kappa^2 c_{k-1}, \ldots\}$ such that the solution $(\vec{d}_k^j, \beta_k^j)$ to the problem $\mathbf{P}^{\mathbf{H_k}}_{\mathbf{c_k^j}, \mathbf{A_k^j}}(\vec{\mathbf{u}}_\mathbf{k})$ satisfies

$$t^{H_k}_{c_k^j, A_k^j}(\vec{u}_k) \leq 0. \tag{3.4}$$

4. Let α_k^j be the largest number from the set $\{\alpha \in \{1, \eta, \eta^2, \ldots\} : \alpha > \tau_k^j\}$ such that

$$\vec{u}_{k+1}^j = \vec{u}_k + \alpha_k^j \vec{d}_k^j$$

satisfies the relation

$$\bar{F}_{c_k^j}(\vec{u}_{k+1}^j) - \bar{F}_{c_k^j}(\vec{u}_k) \leq \gamma \alpha_k^j \sigma^{H_k}_{c_k^j, A_k^j}(\vec{u}_k). \tag{3.5}$$

If no such α_k^j exists, let $\tau_k^{j+1} = 0.5\tau_k^j$, $\xi_k^{j+1} = 0.5\xi_k^j$, increase j by one and return to Step 3.

Otherwise let $\alpha_k = \alpha_k^j$, $\vec{d}_k = \vec{d}_k^j$, $c_k = c_k^j$, $A_k = A_k^j$ and $\vec{u}_{k+1} = \vec{u}_{k+1}^j$. Set $\xi_{k+1} = \xi_k$, $\tau_{k+1} = \tau_k$ if $\xi_k^j = \xi_k$ and $\xi_{k+1} = 0.5\xi_k$, $\tau_{k+1} = 0.5\tau_k$ otherwise. Increase k by one, go to Step 2.

In order to analyse the feasibility of Steps 3–4 we need a new constraint qualification.

$(\mathbf{CQ^N})$: For each $\vec{u} \in \vec{\mathcal{U}}$ we have $\mathcal{F}(\vec{u}) \neq \emptyset$ and, in the case $E \neq \emptyset$,

$$0 \in \text{interior}\,[\mathcal{E}(\vec{u})]\,,$$

where

$$\mathcal{E}(\vec{u}) := \left\{ \left\{ \left\langle \nabla \bar{h}_i^1(\vec{u}), \vec{d} \right\rangle \right\}_{i \in E} : \vec{d} \in \mathcal{F}(\vec{u}) \right\},$$

and

$$\mathcal{F}(\vec{u}) := \left\{ \vec{d} \in \vec{\mathcal{U}} - \vec{u} : \max_{j \in I} \left[\min\left[0, \bar{h}_j^2(\vec{u})\right] + \left\langle \nabla \bar{h}_j^2(\vec{u}), \vec{d} \right\rangle \right] < 0\,, \right.$$
$$\left. \max_{t \in T} \left[\min\left[0, \bar{q}(\vec{u})(t)\right] + \left\langle \nabla \bar{q}(\vec{u})(t), \vec{d} \right\rangle \right] < 0 \right\}.$$

Notice that $\{u^N : u^N \in \mathcal{U}\} \subset \mathcal{U}$. It follows that there can exist u^N which satisfies $(\mathbf{CQ})$ but $\vec{u}^N$ does not fulfill $(\mathbf{CQ^N})$.

Proposition 3.1 *Assume that hypotheses* **(H1)**-**(H3)**, **(BH$^{\mathbf{N}}$)** *and* **(CQ$^{\mathbf{N}}$)** *are satisfied. Then for any* $\vec{u}_k \in \vec{\mathcal{U}}$

(i) *if* $M(\vec{u}_k) = 0$ *then either there exist a finite* l *and* $\alpha_k^l > \tau_k^l > 0$, $c_k^l = c_{k-1}$ *such that (3.4)-(3.5) are satisfied with* $j = l$, *or* $\langle \bar{F}_0(\vec{u}_k), \vec{d}_k \rangle + c_{k-1}\beta_k = 0$ *(where* $(\vec{d}_k, \beta_k)$ *is the solution to the problem* $\mathbf{P}^{\mathbf{H_k}}_{\mathbf{c_{k-1}},\mathbf{R_{\epsilon,u_k}}}(\vec{\mathbf{u}}_\mathbf{k})$*) and* Algorithm 4 *cycles between Steps 3-4,*

(ii) *if* $M(\vec{u}_k) > 0$ *then there exist a finite* l, c_k^l *and* $\alpha_k^l > \tau_k^l > 0$ *such that (3.4)-(3.5) are satisfied with* $j = l$.

Proof.

The proof can be carried out in the same way as the proof of *Proposition 4.1.2.* ■

We can prove that *Algorithm 4* possesses global convergence properties. They refer to the following necessary optimality conditions for $\bar{\vec{u}}$ to be a minimizer of the problem **(P$^{\mathbf{N}}$)**:[1]

(NC$^{\mathbf{N}}$): There exist nonnegative α_j^2, $j \in I$, numbers α_i^1, $i \in E$ and nonnegative measure ν with the support $R_{0,\bar{\vec{u}}}$, not all zero and such that

$$\min_{\vec{d} \in \vec{\mathcal{U}} - \bar{\vec{u}}} \Bigg[\left\langle \nabla \bar{F}_0(\bar{\vec{u}}), \vec{d} \right\rangle + \sum_{i \in E} \alpha_i^1 \left\langle \nabla \bar{h}_i^1(\bar{\vec{u}}), \vec{d} \right\rangle + \sum_{j \in I} \alpha_j^2 \left\langle \nabla \bar{h}_j^2(\bar{\vec{u}}), \vec{d} \right\rangle + \int_T \left\langle \nabla \bar{q}(\bar{\vec{u}})(t), \vec{d} \right\rangle \nu(dt) \Bigg] = 0$$

and $\alpha_j^2 = 0 \ \forall j \notin I_{0,\bar{\vec{u}}} := I_{0,\bar{u}^N}$.

Theorem 3.1 *Assume that the data for* **(P$^{\mathbf{N}}$)** *satisfy hypotheses* **(H1)**-**(H3)**, **(BH$^{\mathbf{N}}$)** *and* **(CQ$^{\mathbf{N}}$)**. *Let* $\{\vec{u}_k\}$ *be a sequence of control functions generated by* Algorithm 4 *and let* $\{c_k\}$ *be the corresponding sequence of penalty parameters. Then*

(i) $\{c_k\}$ *is bounded,*

(ii)

$$\lim_{k \to \infty} \sigma^{H_k}_{c_k, R_{\epsilon, a_k}}(\vec{u}_k) = 0, \quad \lim_{k \to \infty} M(\vec{u}_k) = 0.$$

(iii) *Any limit point of* $\{\vec{u}_k\}$ *satisfies extremality conditions* **(NC$^{\mathbf{N}}$)**.

[1] Necessary conditions **(NC$^{\mathbf{N}}$)** follow directly from (3.6.30). These conditions are more suitable for the semi-infinite programming problem **(P$^{\mathbf{N}}$)** than the conditions **(NC)** applied directly to $\bar{u}^N$.

Proof.

The proof of this theorem is essentially the same as the proofs of *Theorems 3.5.1, 4.1.1* if we take into account the lemma stated below. ∎

Lemma 3.1 *Assume* **(H1)**, **(H2)** *and* $(\mathbf{CQ^N})$. *Then, for any convergent sequence* $\{\vec{u}_k\} \subset \vec{\mathcal{U}}$, *there exist finite* $\bar{k}$, $K_1 > 0$, $K_2 > 0$ *and* $\vec{v} \in \vec{\mathcal{U}}$ *such that for any* $\vec{u}_k$, $k \geq \bar{k}$:

$$\begin{aligned}
\max_{i\in E}\left|\bar{h}_i^1(\vec{u}_k) + \langle \nabla \bar{h}_i^1(\vec{u}_k), \vec{v} - \vec{u}_k\rangle\right| &\leq \max_{i\in E}\left|\bar{h}_i^1(\vec{u}_k)\right| \\
&\quad -K_1 \max_{i\in E}\left|\bar{h}_i^1(\vec{u}_k)\right| \\
\max_{j\in I}\left[\bar{h}_j^2(\vec{u}_k) + \langle \nabla \bar{h}_j^2(\vec{u}_k), \vec{v} - \vec{u}_k\rangle\right] &\leq \max_{j\in I}\bar{h}_j^2(\vec{u}_k) \\
&\quad -K_1 \max_{i\in E}\left|\bar{h}_i^1(\vec{u}_k)\right| \\
\max_{t\in T}\left[\bar{q}(\vec{u}_k)(t) + \langle \nabla \bar{q}(\vec{u}_k)(t), \vec{v} - \vec{u}_k\rangle\right] &\leq \max_{t\in T}\bar{q}(\vec{u}_k)(t) \\
&\quad -K_1 \max_{i\in E}\left|\bar{h}_i^1(\vec{u}_k)\right| \\
\text{and} \quad \|\vec{v} - \vec{u}_k\| &\leq K_2 \max_{i\in E}\left|\bar{h}_i^1(\vec{u}_k)\right|.
\end{aligned}$$

Proof.

The proof of the lemma can be obtained in the same way as the proof of *Lemma 3.6.1.* ∎

We have shown that the method for updating the penalty parameter based on the function $t^H_{c,A}(\vec{u})$ guarantees boundedness of the sequence $\{c_k\}$ under the conditions weaker than those known in the literature.[2] Moreover, the method does not require knowledge of the Lagrange multipliers associated with the subproblem $\mathbf{P}^{\mathbf{H}}_{\mathbf{c},\mathbf{A}}(\vec{\mathbf{u}})$. In the next section we will show that *Algorithm 4* is an SQP type algorithm thus the new method for updating penalty parameters in SQP algorithms is provided.

Frequently, in this chapter, we will consider *Algorithm 4* with the following modifications

(M1): Perform, in Step 3, the additional operations after checking condition (3.4):

> If $\beta_k^j \neq 0$, find $\hat{c}_k^j > \omega c_k^j$ $(\omega > 1)$, $(\hat{\vec{d}}_k^j, \hat{\beta}_k^j)$—the solution to the subproblem $\mathbf{P}^{\mathbf{H_k}}_{\hat{\mathbf{c}}_\mathbf{k}^\mathbf{j},\mathbf{A}_\mathbf{k}^\mathbf{j}}(\vec{\mathbf{u}}_\mathbf{k})$ such that $t^{H_k}_{\hat{c}_k^j,A_k^j}(\vec{u}_k) \leq 0$. Substitute $\hat{c}_k^j$ for c_k^j, $\hat{\vec{d}}_k^j$ for $\vec{d}_k^j$ and $\hat{\beta}_k^j$ for β_j^k.

---

[2] Standard assumption is that gradients of active constraints at a local solution of the problem $(\mathbf{P^N})$ are linearly independent—[12], [41] (and references therein). Our constraint qualification $(\mathbf{CQ^N})$ does not postulate that.

(M2): Define, in Step 4, $\vec{u}^j_{k+1}$ as follows

$$\vec{u}^j_{k+1} = \vec{u}_k + \alpha^j_k \vec{d}^j_k + (\alpha^j_k)^2 \vec{\tilde{d}}^j_k$$

where $\vec{\tilde{d}}^j_k$ is the solution to the second order correction (SOC) subproblem:

$$\min_{\vec{d} \in \mathcal{U} - \vec{u}_k} \left\| \vec{d} \right\|^2$$

subject to the constraints

$$\begin{aligned} \bar{h}^1_i(\vec{u}_k + \vec{d}^j_k) + \left\langle \nabla \bar{h}^1_i(\vec{u}_k), \vec{d} \right\rangle &= 0 \quad \forall i \in E(\vec{u}_k) \\ \bar{h}^2_j(\vec{u}_k + \vec{d}^j_k) + \left\langle \nabla \bar{h}^2_j(\vec{u}_k), \vec{d} \right\rangle &= 0 \quad \forall j \in I(\vec{u}_k) \\ \bar{q}(\vec{u}_k + \vec{d}^j_k)(t) + \left\langle \nabla \bar{h}^2_j(\vec{u}_k), \vec{d} \right\rangle &= 0 \quad \forall t \in T(\vec{u}_k), \end{aligned}$$

where $E(\vec{u}_k) \subset E$, $I(\vec{u}_k) \subset I$ and $T(\vec{u}_k) \subset A^j_k$ are indices of active constraints in the subproblem $\mathbf{P}^{\mathbf{H_k}}_{\mathbf{c}^j_k, \mathbf{A}^j_k}(\vec{\mathbf{u}}_k)$. If a solution to the SOC subproblem does not exist assume $\vec{\tilde{d}}^j_k = 0$.

It is straightforward to show that modifications **(M1)**, **(M2)** do not impair global convergence properties of *Algorithm 4*.

Remark 3.1 *The operation described in* **(M1)** *is needed to guarantee that*

$$c_k > c^{min}_k, \tag{3.6}$$

where $c^{min}_k = \arg\min\{c \geq c_{k-1} : t^{H_k}_{c, A_k}(\vec{u}_k) \leq 0\}$. *The condition (3.6) is required to prove that* Algorithm 4, *when applied to NLP problems, generates* $\{\vec{u}_k\}$ *which is superlinearly convergent to a stationary control. In practice it is very unlikely that at some iteration we have* $c^j_k \leq c^{min}_k$ *so the operation is rarely needed. The choice* $\tilde{c}^j_k = 1.1 c^j_k$ *usually guarantees* $t^{H_k}_{\tilde{c}^j_k, A^j_k}(\vec{u}_k) \leq 0$.

The second order correction which was used in the first order method (c.f. (§4.1.1)) to increase its efficiency plays the same role here. Furthermore, if a) gradients of constraints in $\mathbf{P}^{\mathbf{H_k}}_{\mathbf{c}^j_k, \mathbf{A}^j_k}(\vec{\mathbf{u}}_k)$ are linearly independent, b) $\vec{\mathcal{U}} = \mathcal{R}^{mN}$, c) H_k is a 'good' approximation (the meaning of 'good' is explained in the next section) to the Hessian of the Lagrangian, it also guarantees that close to a local solution $\alpha_k = 1$ and the sequence $\{\vec{u}_k\}$ is superlinearly convergent to a stationary point which satisfies the second order sufficiency conditions (see, e.g. [12]).

4 Bounding the Number of Constraints

In this section we consider an important issue of bounding the number of constraints in $\mathbf{P}^{\mathbf{H}}_{\mathbf{c},\mathbf{A}}(\vec{\mathbf{u}})$. This problem was only partially solved in *Algorithm 4*. We remind that in *Algorithm 4* instead of using all constraints generated by $R_{\varepsilon,\vec{u}_k}$ we use only those from its finite subset A_k. In Chapter 4 we have showed that this approach works well in the feasible directions algorithm. We have demonstrated that even for problems, in which $R_{0,\vec{u}}$ amounts to a large part of T and the control discretization is equal to $N = 1000$, A_k contain no more than few points.

Assume, for the simplicity of presentation, that there is only one active arc of $q(x^{\bar{u}^N}(t))$, which is defined by time t_{en}—the entry time and time t_{ex}—the exit time ($t_{ex} > t_{en}$), and that constraints $b_- \leq u^N(t) \leq b_+$ are not active on $[t_{en}, t_{ex}]$. Furthermore, we assume that there is only one control function. At the active arc we have

$$q(x^{\bar{u}^N}(t)) = 0 \quad \forall t \in [t_{en}, t_{ex}]. \tag{4.1}$$

If we assume that q is sufficiently smooth then we can differentiate q along the $x^{\bar{u}^N}$ trajectory to get the equation

$$S(x^{\bar{u}^N}(t), \bar{u}^N(t), t) = \frac{d^r}{dt^r}\left[q(x^{\bar{u}^N}(t))\right] = 0 \quad \forall t \in [t_{en}, t_{ex}], \tag{4.2}$$

where r is the smallest positive integer number (assuming that it exists) for which the function S depends explicitly on $\bar{u}^N$.

If S is continuously differentiable with respect to all its arguments and

$$\frac{\partial}{\partial u} S(x, u, t) \neq 0 \tag{4.3}$$

along the active arc then from the Implicit Function Theorem, we can express u as a differentiable function of x. That implies that if

$$[(i-1)/N, i/N] \cap [t_{en}, t_{ex}] = C_i \neq \emptyset$$

for some $i = 1, \ldots, N$, then, since $\bar{u}^N$ is constant on C_i, only one point from C_i should be sufficient to determine $\bar{u}^N$ on C_i. We denote the set of these $i \in \{1, 2, \ldots, N\}$ for which $C_i \neq \emptyset$ by I_C.

The derivations presented above are based on assumption (4.3). This assumption is very often imposed on optimal control problems with state inequality constraints, see for example [67] and [129] where the similar assumption was imposed on mixed state–control constraints. Moreover, the transformation of (4.1) to (4.2) is used, in explicit or inexplicit form, in various computational methods for control problems with state constraints ([76],[81],[27]).

Having outlined this quasi–analytical approach to control problems with state constraints we would like to know how it could be incorporated into

Algorithm 4. We can anticipate, from the above considerations, that only one point from C_i should be included in A_k if some condition similar to (4.3) is met.[3] We will show that, alternatively, the set A_k can be constructed out of points $t_i \in C_i$ such that $\nabla \bar{q}(\vec{u})(t_i)$ are linearly independent.

Assume that (4.3) holds for $r = 1$. This implies that

$$q_x(x^{\bar{u}^N}(t_i))f_u(t_i, \bar{u}^N(t_i), x^{\bar{u}^N}(t_i)) \neq 0 \tag{4.4}$$

(c.f. the definition of uniform independence in [37], p. 699). Hypothesize now that the condition analogous to (4.4) holds:

$$\int_{(i-1)/N}^{t_i} f_u(t, x^{\bar{u}^N}(t), \bar{u}^N(t))^T p(t)dt \neq 0, \tag{4.5}$$

where

$$\begin{aligned} p(t_i) &= q_x(x^{\bar{u}^N}(t_i))^T \\ \dot{p}(t) &= -f_x(t, x^{\bar{u}^N}(t), \bar{u}^N(t))^T p(t), \ t \in [(i-1)/N, t_i). \end{aligned}$$

Then vectors $\nabla \bar{q}(\vec{u})(t_i)$, $i \in I_C$ are linearly independent.

In order to prove that consider

$$\sum_{i \in I_C} \alpha^i \nabla \bar{q}(\vec{\bar{u}})(t_i) = 0. \tag{4.6}$$

If we order $\{t_i\}_{i \in I_C}$ from the biggest value to the smallest one—$\{t_{j_i}\}_1^{|I_C|}$ we will obtain from (4.5)–(4.6) and the fact that $\nabla \bar{q}(\vec{\bar{u}})(t) = 0$, for $t > t_i$, $t_i \in I_C$:

$$\alpha^{j_1} = 0 \Rightarrow \alpha^{j_2} = 0 \Rightarrow \ldots \Rightarrow \alpha^{j_{|I_C|}} = 0 \tag{4.7}$$

which completes the proof.

Notice that (4.4) is not prerequisite for (4.5) to hold. Consider the optimal control problem discussed in [57]:

$$\min_u \int_0^3 2x_1(t)dt$$

$$\begin{aligned} \text{s. t. } \dot{x}_1(t) &= x_2(t) \\ \dot{x}_2(t) &= u(t) \\ -2 \leq u(t) &\leq 2 \\ x_1(t) &\geq \alpha, \ t \in [0,3]. \end{aligned}$$

[3] If there are more than one control function and the function S in (4.2) depends on all of them, more constraints can be needed to determine $\bar{u}^N$ on C_i. Note that this case is not covered by assumption (4.3) discussed above since we assume that q is a scalar function.

For $x_1(0) = 2$, $x_2(0) = 0$ and $\alpha \in (-2.5, 0)$ there is an active arc at the solution. Condition (4.4) is not satisfied since (after the horizon normalization):

$$q_x(x^{\bar{u}^N}(t_i)) f_u(t_i, x^{\bar{u}^N}(t_i), \bar{u}^N(t_i)) = \\ [1, 0] \begin{bmatrix} 0 \\ 3 \end{bmatrix} = 0.$$

However, as it is easy to check, (4.5) holds.

Conditions under which the problem ($\mathbf{P^N}$) can be reduced to a problem with a finite number of constraints, at all local solutions, is the subject of this section.

4.1 Some Remarks on Direction Finding Subproblems

Consider the subproblem $\mathbf{P^H_{c,A}}(\bar{\mathbf{u}})$. It can be restated as

$$\min_{\vec{d} \in \mathcal{U} - \vec{u}, \beta \in \mathcal{R}} \left[\left\langle \nabla \bar{F}_0(\vec{u}), \vec{d} \right\rangle + c\beta + 1/2 (\vec{d})^T H \vec{d} \right]$$

subject to

$$\begin{aligned} \bar{h}^1_{i,1}(\vec{u}) + \left\langle \nabla \bar{h}^1_{i,1}(\vec{u}), \vec{d} \right\rangle &\leq \beta \ \forall i \in E \\ \bar{h}^1_{i,2}(\vec{u}) + \left\langle \nabla \bar{h}^1_{i,2}(\vec{u}), \vec{d} \right\rangle &\leq \beta \ \forall i \in E \\ \bar{h}^2_j(\vec{u}) + \left\langle \nabla \bar{h}^2_j(\vec{u}), \vec{d} \right\rangle &\leq \beta \ \forall j \in I \\ \bar{q}(\vec{u})(t) + \left\langle \nabla \bar{q}(\vec{u})(t), \vec{d} \right\rangle &\leq \beta \ \forall t \in A. \end{aligned}$$

Here,

$$\begin{aligned} h^1_{i,1}(\vec{u}) &:= h^1_i(\vec{u}) \ \forall i \in E \\ h^1_{i,2}(\vec{u}) &:= -h^1_i(\vec{u}) \ \forall i \in E. \end{aligned}$$

If $(\vec{\hat{d}}, \hat{\beta} = 0)$ is the solution to $\mathbf{P^H_{c,A}}(\vec{\mathbf{u}})$, then $\vec{\hat{d}}$ solves also the subproblem $\mathbf{P^{SQP}_{A,H}}(\bar{\mathbf{u}})$ defined as follows

$$\min_{\vec{d} \in \mathcal{U} - \vec{u}} \left[\left\langle \nabla \bar{F}_0(\vec{u}), \vec{d} \right\rangle + 1/2 (\vec{d})^T H \vec{d} \right]$$

subject to

$$\begin{aligned} \bar{h}^1_i(\vec{u}) + \left\langle \nabla \bar{h}^1_i(\vec{u}), \vec{d} \right\rangle &= 0 \ \forall i \in E \\ \bar{h}^2_j(\vec{u}) + \left\langle \nabla \bar{h}^2_j(\vec{u}), \vec{d} \right\rangle &\leq 0 \ \forall j \in I \\ \bar{q}(\vec{u})(t) + \left\langle \nabla \bar{q}(\vec{u})(t), \vec{d} \right\rangle &\leq 0 \ \forall t \in A \end{aligned}$$

(otherwise $(\vec{d}, 0)$ would provide lower value for $\mathbf{P}^{\mathbf{H}}_{\mathbf{c},\mathbf{A}}(\vec{\mathbf{u}})$, where $\vec{d}$ is the solution to $\mathbf{P}^{\mathbf{SQP}}_{\mathbf{A},\mathbf{H}}(\vec{\mathbf{u}})$). Notice that c does not appear in the formulation of the subproblem $\mathbf{P}^{\mathbf{SQP}}_{\mathbf{A},\mathbf{H}}(\vec{\mathbf{u}})$.

The subproblem $\mathbf{P}^{\mathbf{SQP}}_{\mathbf{A},\mathbf{H}}(\vec{\mathbf{u}})$ plays an important role in many SQP algorithms. According to [79] nomenclature these are methods based on inequality–constrained QP subproblems. The role of $\mathbf{P}^{\mathbf{SQP}}_{\mathbf{A},\mathbf{H}}(\vec{\mathbf{u}})$ in these algorithms is twofold: a) to calculate a direction of descent for the *merit function* (such as the function $\tilde{F}_c$ in (1.1)) used in the SQP algorithm; b) to predict the Lagrange multipliers used in the *Lagrangian function* whose Hessian, or its approximation, defines H ([79]). We will show that the subproblem $\mathbf{P}^{\mathbf{H}}_{\mathbf{c},\mathbf{A}}(\vec{\mathbf{u}})$ is closely related to $\mathbf{P}^{\mathbf{SQP}}_{\mathbf{A},\mathbf{H}}(\vec{\mathbf{u}})$.

The solution $(\hat{\vec{d}}, \hat{\beta})$ to $\mathbf{P}^{\mathbf{H}}_{\mathbf{c},\mathbf{A}}(\vec{\mathbf{u}})$ satisfies the set of equations[4]

$$\begin{aligned}
\bar{h}^1_{i,1}(\vec{u}) + \left\langle \nabla \bar{h}^1_{i,1}(\vec{u}), \hat{\vec{d}} \right\rangle &= \hat{\beta} \ \ \forall i \in E^+(\vec{u}) \\
\bar{h}^1_{i,2}(\vec{u}) + \left\langle \nabla \bar{h}^1_{i,2}(\vec{u}), \hat{\vec{d}} \right\rangle &= \hat{\beta} \ \ \forall i \in E^-(\vec{u}) \\
\bar{q}(\vec{u})(t) + \left\langle \nabla \bar{q}(\vec{u})(t), \hat{\vec{d}} \right\rangle &= \hat{\beta} \ \ \forall t \in T(\vec{u}) \\
\bar{h}^2_j(\vec{u}) + \left\langle \nabla \bar{h}^2_j(\vec{u}), \hat{\vec{d}} \right\rangle &= \hat{\beta} \ \ \forall j \in I(\vec{u}),
\end{aligned}$$

where $E^+(\vec{u}) \subset E$, $E^-(\vec{u}) \subset E$, $I(\vec{u}) \subset I$, $T(\vec{u}) \subset A$. Furthermore, $E^+(\vec{u})$, $E^-(\vec{u})$, $I(\vec{u})$ and $T(\vec{u})$ define the maximum set of constraints which are active at $(\hat{\vec{d}}, \hat{\beta})$ and whose gradients are linearly independent. If $\hat{\beta} \neq 0$ then we have

$$E^+(\vec{u}) \bigcap E^-(\vec{u}) = \emptyset. \tag{4.8}$$

The above equations can be written as

$$\tilde{A} \begin{bmatrix} \hat{\vec{d}} \\ \hat{\beta} \end{bmatrix} =$$

[4] To simplify the presentation in the rest of this section we assume that simple constraints (4.2.3) are not present or have been included in the inequality constraints $\bar{h}^2_j$.

$$\begin{bmatrix} \nabla\bar{h}^1_{i_1,1}(\vec{u})^T, & -1 \\ \vdots & \vdots \\ \nabla\bar{h}^1_{i_{|E^+(\vec{u})|},1}(\vec{u})^T, & -1 \\ \nabla\bar{h}^1_{i_1,2}(\vec{u})^T, & -1 \\ \vdots & \vdots \\ \nabla\bar{h}^1_{i_{|E^-(\vec{u})|},2}(\vec{u})^T, & -1 \\ \nabla\bar{q}(\vec{u})(t_1)^T, & -1 \\ \nabla\bar{q}(\vec{u})(t_2)^T, & -1 \\ \vdots & \vdots \\ \nabla\bar{q}(\vec{u})(t_{|T(\vec{u})|})^T, & -1 \\ \nabla\bar{h}^2_{j_1}(\vec{u})^T, & -1 \\ \vdots & \vdots \\ \nabla\bar{h}^2_{j_{|I(\vec{u})|}}(\vec{u})^T, & -1 \end{bmatrix} \begin{bmatrix} \vec{d} \\ \hat{\beta} \end{bmatrix} = \begin{bmatrix} -\bar{h}^1_{i_1,1}(\vec{u}) \\ \vdots \\ -\bar{h}^1_{i_{|E(\vec{u})|},1}(\vec{u}) \\ -\bar{h}^1_{i_1,2}(\vec{u}) \\ \vdots \\ -\bar{h}^1_{i_{|E(\vec{u})|},2}(\vec{u}) \\ -\bar{q}(\vec{u})(t_1) \\ -\bar{q}(\vec{u})(t_2) \\ \vdots \\ -\bar{q}(\vec{u})(t_{|T(\vec{u})|}) \\ -\bar{h}^2_{j_1}(\vec{u}) \\ \vdots \\ \bar{h}^2_{j_{|I(\vec{u})|}}(\vec{u}) \end{bmatrix} = \bar{a}. \tag{4.9}$$

On the other hand, the solution $\vec{d}$ to the subproblem $\mathbf{P}^{\mathbf{SQP}}_{\mathbf{A},\mathbf{H}}(\vec{\mathbf{u}})$ satisfies the set of equations:

$$\begin{aligned} \bar{h}^1_i(\vec{u}) + \left\langle \nabla\bar{h}^1_i(\vec{u}), \vec{d} \right\rangle &= 0 \ \forall i \in E(\vec{u}) \\ \bar{q}(\vec{u})(t) + \left\langle \nabla\bar{q}(\vec{u})(t), \vec{d} \right\rangle &= 0 \ \forall t \in T(\vec{u}) \\ \bar{h}^2_j(\vec{u}) + \left\langle \nabla\bar{h}^2_j(\vec{u}), \vec{d} \right\rangle &= 0 \ \forall j \in I(\vec{u}). \end{aligned}$$

Here, $E(\vec{u}) \subset E$ and $E(\vec{u})$, $I(\vec{u})$, $T(\vec{u})$ define the maximum set of constraint functions whose gradients are linearly independent. (We use the same set of indices $T(\vec{u})$ and $I(\vec{u})$ to denote active constraints at the solutions to both $\mathbf{P}^{\mathbf{H}}_{\mathbf{c},\mathbf{A}}(\vec{\mathbf{u}})$ and $\mathbf{P}^{\mathbf{SQP}}_{\mathbf{A},\mathbf{H}}(\vec{\mathbf{u}})$ subproblems—the justification for that is given later in the subsection.) We can also state the above equations as follows

$$\bar{A}\vec{d} = \bar{a}. \tag{4.10}$$

There two major differences between subproblems $\mathbf{P}^{\mathbf{H}}_{\mathbf{c},\mathbf{A}}(\vec{\mathbf{u}})$ and $\mathbf{P}^{\mathbf{SQP}}_{\mathbf{A},\mathbf{H}}(\vec{\mathbf{u}})$ in the case when $(\vec{d}, \beta = 0)$ solves $\mathbf{P}^{\mathbf{H}}_{\mathbf{c},\mathbf{A}}(\vec{\mathbf{u}})$.

The Lagrange multipliers corresponding to active constraints at the solution to $\mathbf{P}^{\mathbf{H}}_{\mathbf{c},\mathbf{A}}(\vec{\mathbf{u}})$ are bounded between 0 and c (we have inequality constraints —thus they are positive, moreover the sum of the Lagrange multipliers must be equal to c—this follows from the fact that the derivative of the Lagrangian with respect to β must be equal to zero). On the other hand, we cannot provide any *a priori* bounds on the Lagrange multipliers of the active constraints at the solution to $\mathbf{P}^{\mathbf{SQP}}_{\mathbf{A},\mathbf{H}}(\vec{\mathbf{u}})$.

The other difference is the maximum number of constraints active at the solutions to $\mathbf{P}^{\mathbf{H}}_{\mathbf{c},\mathbf{A}}(\vec{\mathbf{u}})$ (when $\hat{\beta} = 0$) and $\mathbf{P}^{\mathbf{SQP}}_{\mathbf{A},\mathbf{H}}(\vec{\mathbf{u}})$. However, we show below that there can be only one $i \in E$ such that $i \in E^+(\vec{u}) \bigcap E^-(\vec{u})$ which implies that $|E^+(\vec{u}) \bigcup E^-(\vec{u})| \leq |E(\vec{u})| + 1$.

Suppose first that there are more than one $i \in E$ such that $i \in E^+(\vec{u}) \bigcap E^-(\vec{u})$. Build the matrix $\bar{A}$ using $E(\vec{u}) = E^+(\vec{u}) \bigcup E^-(\vec{u})$. It is easy to show that the matrix $\bar{A}$, composed in this way, has the full row rank. Checking the condition for the linear independence of vectors definning $\bar{A}$ and $\tilde{A}$ (c.f. (4.9) and (4.10)) results in the equations

$$\begin{aligned}
\alpha(t) &= 0 \quad \forall t \in T(\vec{u}) \\
\alpha_j^2 &= 0 \quad \forall j \in I(\vec{u}) \\
\alpha_{i,1}^1 &= 0 \quad \forall i \in E^+(\vec{u}) \backslash [E^+(\vec{u}) \bigcap E^-(\vec{u})] \\
\alpha_{i,2}^1 &= 0 \quad \forall i \in E^-(\vec{u}) \backslash [E^+(\vec{u}) \bigcap E^-(\vec{u})] \\
\alpha_{i,1}^1 &= \alpha_{i,2}^1 \quad \forall i \in E^+(\vec{u}) \bigcap E^-(\vec{u}) \\
\sum_{i \in E^+(\vec{u}) \bigcap E^-(\vec{u})} (\alpha_{i,1}^1 + \alpha_{i,2}^1) &= 0.
\end{aligned}$$

Here, $\alpha(t), t \in T(\vec{u})$, for instance, are coefficients corresponding to $\nabla \bar{q}(\vec{u})(t)$, $t \in T(\vec{u})$. The last two equations can be reduced to the equation

$$\sum_{i \in E^+(\vec{u}) \bigcap E^-(\vec{u})} \alpha_{i,1}^1 = 0.$$

Therefore, if $|E^+(\vec{u}) \bigcap E^-(\vec{u})| > 1$ then we could find nonzero $\alpha_{i,1}^1$, $\alpha_{i,2}^1$, $i \in E^+(\vec{u}) \bigcap E^-(\vec{u})$ for which the above set of equations is satisfied. This contradicts the fact that $\bar{A}$ has full row rank.

Let $(\hat{\vec{d}}, \hat{\beta} = 0)$ be the solution to $\mathbf{P}_{\mathbf{c,A}}^{\mathbf{H}}(\bar{\mathbf{u}})$ and let $E^+(\vec{u})$, $E^-(\vec{u})$, $I(\vec{u})$, $T(\vec{u})$ define the maximum set of linearly independent vectors which form equations (4.9). We will show that $E^+(\vec{u}) \bigcup E^-(\vec{u})$, $I(\vec{u})$ and $T(\vec{u})$ describe a maximum set for the subproblem $\mathbf{P}_{\mathbf{A,H}}^{\mathbf{SQP}}(\bar{\mathbf{u}})$, i.e., $E(\vec{u}) = E^+(\vec{u}) \bigcup E^-(\vec{u})$.

Assume the contrary. This means that there exists, say, $\bar{j} \in I$ such that vectors $\nabla \bar{h}_i^1(\vec{u})$, $i \in E(\vec{u})$, $\nabla \bar{h}_j^2(\vec{u})$, $j \in I(\vec{u})$, $\nabla \bar{q}(\vec{u})(t)$, $t \in T(\vec{u})$ and $\nabla \bar{h}_{\bar{j}}^2(\vec{u})$ are linearly independent and define the solution to $\mathbf{P}_{\mathbf{A,H}}^{\mathbf{SQP}}(\bar{\mathbf{u}})$. Vectors $[\nabla \bar{h}_i^1(\vec{u})^T, -1]^T, i \in E(\vec{u})$, $[\nabla \bar{h}_j^2(\vec{u})^T, -1]^T, j \in I(\vec{u})$, $[\nabla \bar{q}(\vec{u})(t)^T, -1]^T$, $t \in T(\vec{u})$ are also linearly independent and

$$\begin{aligned}
\nabla \bar{h}_{\bar{i}}^1(\vec{u}) &= \sum_{t \in T(\vec{u})} \alpha(t) \nabla \bar{q}(\vec{u})(t) + \sum_{j \in I(\vec{u})} \alpha_j^2 \nabla \bar{h}_j^2(\vec{u}) + \\
&\quad \sum_{i \in E(\vec{u})} \alpha_i^1 \nabla \bar{h}_i^1(\vec{u}) + \alpha_{\bar{j}}^2 \nabla \bar{h}_{\bar{j}}^2(\vec{u}) \qquad (4.11) \\
1 &= \sum_{t \in T(\vec{u})} \alpha(t) + \sum_{j \in I(\vec{u})} \alpha_j^2 + \sum_{i \in E(\vec{u})} \alpha_i^1 + \alpha_{\bar{j}}^2, \qquad (4.12)
\end{aligned}$$

where $\bar{i} \in E^+(\vec{u}) \bigcap E^-(\vec{u})$ because otherwise $E^+(\vec{u})$, $E^-(\vec{u})$, $I(\vec{u})$, $T(\vec{u})$ and $\bar{j}$ would define the maximum set. Since vectors, which appear on the right-hand side of (4.11), are linearly independent and $\bar{i} \in E(\vec{u})$ we must have $\alpha_{\bar{i}}^1 = -1$ and all other coefficients in (4.11) equal zero. But this contradicts (4.12). Therefore, we have shown that we can write $E(\vec{u}) = E^+(\vec{u}) \bigcup E^-(\vec{u})$ and that together with (4.8) enable us to introduce the following notation which plays an important role in this section.

Notation 4.1 *We denote by $T(\vec{u})$, $E^+(\vec{u})$, $E^-(\vec{u})$ ($E(\vec{u})$), $I(\vec{u})$ indices of these state, equality and inequality constraints whose gradients form a maximum set of linearly independent vectors among gradients of all active constraints at the solution to the subproblem $\mathbf{P}^{\mathbf{H}}_{\mathbf{c,A}}(\vec{\mathbf{u}})$ ($\mathbf{P}^{\mathbf{SQP}}_{\mathbf{A,H}}(\vec{\mathbf{u}})$).*

4.2 The Nonlinear Programming Problem

Assume that the following conditions are satisfied at $\bar{\vec{u}}$ which a stationary point for the problem ($\mathbf{P^N}$).

(H4) Let $\nabla\bar{q}(\bar{\vec{u}})(t)$, $t \in T(\bar{\vec{u}})$, $\nabla\bar{h}_i^1(\bar{\vec{u}})$, $i \in E(\bar{\vec{u}})$, $\nabla\bar{h}_j^2(\bar{\vec{u}})$, $j \in I(\bar{\vec{u}})$ be such that there exist the Lagrange multipliers $\bar{\lambda}_i^1$, $i \in E(\bar{\vec{u}})$, $\bar{\lambda}_j^2$, $j \in I(\bar{\vec{u}})$, $\bar{\lambda}(t)$, $t \in T(\bar{\vec{u}})$ satisfying

$$\nabla\bar{F}_0(\bar{\vec{u}}) + \sum_{i\in E(\bar{\vec{u}})} \bar{\lambda}_i^1\nabla\bar{h}_i^1(\bar{\vec{u}}) + \sum_{t\in T(\bar{\vec{u}})} \bar{\lambda}(t)\nabla\bar{q}(\bar{\vec{u}})(t) + \sum_{i\in I(\bar{\vec{u}})} \bar{\lambda}_i^2\nabla\bar{h}_j^2(\bar{\vec{u}}) = 0. \quad (4.13)$$

Then

(i)

$$\bar{\lambda}_i^1 \neq 0 \ \ \forall i \in E(\bar{\vec{u}}),\ \bar{\lambda}_j^2 > 0 \ \ \forall j \in I(\bar{\vec{u}}),\quad \bar{\lambda}(t) > 0 \ \ \forall t \in T(\bar{\vec{u}}). \quad (4.14)$$

(ii) If $\bar{H}$ is defined by

$$\bar{H} = \nabla_{uu}^2\bar{F}_0(\bar{\vec{u}}) + \sum_{i\in E(\bar{\vec{u}})} \bar{\lambda}_i^1\nabla_{uu}^2\bar{h}_i^1(\bar{\vec{u}}) + \sum_{t\in T(\bar{\vec{u}})} \bar{\lambda}(t)\nabla_{uu}^2\bar{q}(\bar{\vec{u}})(t) + \sum_{j\in I(\bar{\vec{u}})} \bar{\lambda}_j^2\nabla_{uu}^2\bar{h}_j^2(\bar{\vec{u}})$$

then

$$(\vec{d})^T\bar{H}\vec{d} > 0 \quad (4.15)$$

for all $\vec{d} \neq 0$ such that

$$\left\langle \nabla \bar{h}_i^1(\vec{\bar{u}}), \vec{d} \right\rangle = 0 \;\; \forall i \in E(\vec{\bar{u}}) \tag{4.16}$$

$$\left\langle \nabla \bar{q}(\vec{\bar{u}})(t), \vec{d} \right\rangle = 0 \;\; \forall t \in T(\vec{\bar{u}}) \tag{4.17}$$

$$\left\langle \nabla \bar{h}_j^2(\vec{\bar{u}}), \vec{d} \right\rangle = 0 \;\; \forall j \in I(\vec{\bar{u}}), \tag{4.18}$$

where all second order derivatives are assumed to exist and be continuous. Moreover, $\nabla^2_{uu}\bar{q}(\vec{u})(\cdot)$ is continuous.

(4.14) is the strict complementarity condition which is usually required if a superlinear rate of convergence is to be guaranteed for an SQP type algorithm.

The assumption (**H4**) emphasizes the fact that the optimality conditions can only be checked for a maximum set of linearly independent vectors $\nabla \bar{h}_i^1(\vec{\bar{u}})$, $i \in E(\vec{\bar{u}})$, $\nabla \bar{h}_j^2(\vec{\bar{u}})$, $j \in I(\vec{\bar{u}})$, $\nabla \bar{q}(\vec{\bar{u}})(t)$, $t \in T(\vec{\bar{u}})$. This follows from the fact that the solution to $\mathbf{P}^{\mathbf{SQP}}_{\mathbf{A_k},\mathbf{H_k}}(\vec{\mathbf{u}}_\mathbf{k})$ is uniquely determined by any set of vectors as defined in *Notation 4.1*.

Consider the problem

$$\min_{\vec{d}} \left[\left\langle \nabla \bar{F}_0(\vec{u}_k), \vec{d} \right\rangle + 1/2(\vec{d})^T H_k \vec{d} \right]$$

$$\text{s. t.} \quad A_k \vec{d} = a_k,$$

where A_k and a_k are defined by all active constraints at a solution to $\mathbf{P}^{\mathbf{SQP}}_{\mathbf{A_k},\mathbf{H_k}}(\vec{\mathbf{u}}_\mathbf{k})$. (The use of $\mathbf{P}^{\mathbf{SQP}}_{\mathbf{A_k}.\mathbf{H_k}}(\vec{\mathbf{u}})$ instead of $\mathbf{P}^{\mathbf{H_k}}_{\mathbf{c_k}.\mathbf{A_k}}(\vec{\mathbf{u}}_\mathbf{k})$ is justified later.) Notice that any maximum set of linearly independent vectors as stated in *Notation 4.1* can be used to define the solution $\vec{d}_k$ and the corresponding set of the Lagrange multipliers (which depends on the choice of the maximum set). But the solution $\vec{d}_k$ is unique since $\mathbf{P}^{\mathbf{SQP}}_{\mathbf{A_k},\mathbf{H_k}}(\vec{\mathbf{u}}_\mathbf{k})$ has a strictly convex objective function.

The assumption (**H4**) is related to sufficiency conditions for the problem $(\mathbf{P_{NLP}}(\vec{\bar{\mathbf{u}}}))$:

$$\min_{\vec{u}} \bar{F}_0(\vec{u}) \tag{4.19}$$

subject to the constraints

$$\bar{h}_i^1(\vec{u}) = 0 \;\; \forall i \in E(\vec{\bar{u}}) \tag{4.20}$$

$$\bar{q}(\vec{u})(t) \leq 0 \;\; \forall t \in T(\vec{\bar{u}}) \tag{4.21}$$

$$\bar{h}_j^2(\vec{u}) \leq 0 \;\; \forall j \in I(\vec{\bar{u}}). \tag{4.22}$$

We recall that $\vec{\bar{u}}$ is a strong local minimum for the problem $(\mathbf{P_{NLP}}(\vec{\bar{\mathbf{u}}}))$ if there exists $\gamma_1 > 0$ such that

$$\bar{F}_0(\vec{u}) \geq \bar{F}_0(\vec{\bar{u}}) + \gamma_1 \left\| \vec{u} - \vec{\bar{u}} \right\|^2$$

for all $\vec{u} \in \mathcal{B}(\bar{\vec{u}}, \varepsilon)$ satisfying (4.20)–(4.22). Here, $\mathcal{B}(\bar{\vec{u}}, \varepsilon)$ is an open ball with the center $\bar{\vec{u}}$ and the radius $\varepsilon > 0$.

The importance of the assumption (**H4**) is highlighted in the following proposition.

Proposition 4.1 *(i) Assume that* (**H4**) *is satisfied at* $\bar{\vec{u}}$*. Then* $\bar{\vec{u}}$ *is a strong local minimum point for the problem* ($\mathbf{P_{NLP}}(\bar{\vec{\mathbf{u}}})$).

(ii) Suppose that $\bar{\vec{u}}$ *is a strong local minimum point for the problem* ($\mathbf{P^N}$) *in the sense that there exists* $\gamma > 0$ *such that* $\bar{F}_0(\vec{u}) \geq \bar{F}_0(\bar{\vec{u}}) + \gamma\|\vec{u} - \bar{\vec{u}}\|^2$ *for all* $\vec{u}$*, from some neighbourhood of* $\bar{\vec{u}}$*, satisfying the constraints of* ($\mathbf{P^N}$)*. Then* (ii) *of* (**H4**) *is satisfied.*

(iii) Assume that (**H4**) *holds. Then* $\bar{\vec{u}}$ *is a strong local minimum point for the penalty function*

$$\bar{F}_c^R(\vec{u}) = \bar{F}_0(\vec{u}) + c \max \left[0, \max_{i \in E(\bar{\vec{u}})} \left| \bar{h}_i^1(\vec{u}) \right|, \right. \\ \left. \max_{t \in T(\bar{\vec{u}})} \bar{q}(\vec{u})(t), \max_{j \in I(\bar{\vec{u}})} \bar{h}_j^2(\vec{u}) \right] \tag{4.23}$$

if c *satisfies*

$$c \geq \sum_{t \in T(\bar{\vec{u}})} \bar{\lambda}(t) + \sum_{j \in I(\bar{\vec{u}})} \bar{\lambda}_j^2 + \sum_{i \in E(\bar{\vec{u}})} |\bar{\lambda}_i^1|.$$

Proof.

(i) is a well-known result, see, for instance, [70].

(ii) Using the same arguments as in [70] (pp. 226–234) we can show that there exists $\gamma > 0$ such that for all $\vec{d} \neq 0$ satisfying $\langle \nabla \bar{h}_i^1(\bar{\vec{u}}), \vec{d} \rangle = 0 \; \forall i \in E$, $\langle \nabla \bar{q}(\bar{\vec{u}})(t), \vec{d} \rangle = 0 \; \forall t \in R_{0,\bar{\vec{u}}}$, $\langle \nabla \bar{h}_j^2(\bar{\vec{u}}), \vec{d} \rangle = 0 \; \forall j \in I_{0,\bar{\vec{u}}}$ we have

$$(\vec{d})^T \left[\nabla_{uu}^2 \bar{F}_0^{\gamma}(\bar{\vec{u}}) + \sum_{i \in E(\bar{\vec{u}})} \bar{\lambda}_i^1 \nabla_{uu}^2 \bar{h}_i^1(\bar{\vec{u}}) + \right. \\ \left. \sum_{t \in T(\bar{\vec{u}})} \bar{\lambda}(t) \nabla_{uu}^2 \bar{q}(\bar{\vec{u}})(t) + \sum_{j \in I(\bar{\vec{u}})} \bar{\lambda}_j^2 \nabla_{uu}^2 \bar{h}_j^2(\bar{\vec{u}}) \right] \vec{d} \geq 0, \tag{4.24}$$

where $\bar{F}_0^{\gamma}(\vec{u}) = \bar{F}_0(\vec{u}) + \gamma\|\vec{u} - \bar{\vec{u}}\|^2$. From the definition of $E(\bar{\vec{u}})$, $T(\bar{\vec{u}})$, $I(\bar{\vec{u}})$ and (4.24) we conclude that (4.15) holds.

(iii) follows from Theorem 4.6 in [29]. ∎

Proposition 4.1 (iii) implies that there exist $\gamma_2 > 0$ and $\varepsilon > 0$ such that

$$\bar{F}_c^R(\vec{u}) \geq \bar{F}_c^R(\bar{\vec{u}}) + \gamma_2 \left\| \vec{u} - \bar{\vec{u}} \right\|^2 \quad \forall \vec{u} \in \mathcal{B}(\bar{\vec{u}}, \varepsilon). \tag{4.25}$$

This property is extensively used in the proof of the theorem stated in this subsection.

The theorem examines the implications of the fact that under hypothesis (**H4**) the point $\bar{\vec{u}}$ is a strong local minimum for the problem ($\mathbf{P_{NLP}}(\bar{\vec{\mathbf{u}}})$). One can expect that, if we are close to such a point, the algorithm modified for the problem ($\mathbf{P_{NLP}}(\bar{\vec{\mathbf{u}}})$) finds $\bar{\vec{u}}$ because, as we will show, $\bar{u}^N$ is a point of attraction for a sequence $\{\vec{u}_k\}$ generated by *Algorithm 4* adapted for the problem ($\mathbf{P_{NLP}}(\bar{\vec{\mathbf{u}}})$).

Before presenting the theorem we need to prove that *Algorithm 4* becomes an SQP algorithm when applied to the problem ($\mathbf{P_{NLP}}(\bar{\vec{\mathbf{u}}})$).

Assume that, at some neighbourhood of $\bar{\vec{u}}$, $\vec{d}_k$ is determined by solving the problem

$$\min_{(\vec{d},\beta)} \left[\left\langle \nabla \bar{F}_0(\vec{u}_k), \vec{d}\right\rangle + c_k\beta + 1/2(\vec{d})^T H_k \vec{d}\right] \tag{4.26}$$

subject to the constraints

$$\tilde{A}_k \begin{bmatrix} \vec{d} \\ \beta \end{bmatrix} = \tilde{a}_k, \tag{4.27}$$

where $\tilde{A}_k$, $\tilde{a}_k$ are defined as in (4.9) with the exception that the sets of indices $E^+(\vec{u})$, $E^-(\vec{u})$, $I(\vec{u})$ and $T(\vec{u})$ are replaced by $E^+(\bar{\vec{u}})$, $E^-(\bar{\vec{u}})$, $I(\bar{\vec{u}})$ and $T(\bar{\vec{u}})$. Furthermore, suppose that the following condition holds.

(**H5**) There exists $\hat{k}$ such that for all $k \geq \hat{k}$

$$\Big\{\vec{d} \in \vec{\mathcal{U}} - \vec{u}_k : \quad \begin{aligned} &\bar{h}_i^1(\vec{u}_k) + \left\langle \nabla \bar{h}_i^1(\vec{u}_k), \vec{d}\right\rangle = 0 \ \ \forall i \in E, \\ &\bar{h}_j^2(\vec{u}_k) + \left\langle \nabla \bar{h}_j^2(\vec{u}_k), \vec{d}\right\rangle \leq 0 \ \ \forall j \in I, \\ &\bar{q}(\vec{u}_k)(t) + \left\langle \nabla \bar{q}(\vec{u}_k)(t), \vec{d}\right\rangle \leq 0 \ \ \forall t \in R_{\varepsilon,\vec{u}_k}\Big\} \neq \emptyset. \end{aligned}$$

(**H5**) is an extension of the widely imposed condition on nonlinear programming problems: close to a local solution direction finding subproblems defined by linear approximations of active constraints have feasible points $(\vec{d}, \beta)$ such that $\beta = 0$.

Under the hypothesis (**H5**) we can prove the following lemma.

Lemma 4.1 *Suppose that* Algorithm 4, *with the modification* (**M1**), *is applied to the problem* ($\mathbf{P_{NLP}}(\bar{\vec{\mathbf{u}}})$) *defined by* $E(\bar{\vec{u}})$, $I(\bar{\vec{u}})$, $T(\bar{\vec{u}})$ *for which (4.13) holds. Assume also that*

(i) (**H1**)-(**H3**), (**H5**), ($\mathbf{BH^N}$) *and* ($\mathbf{CQ^N}$) *are satisfied,*

(ii) there exists a neighbourhood $\mathcal{N}$ of $\bar{\vec{u}}$ such that if $\vec{u}_k \in \mathcal{N}$ then the solution to the subproblem $\mathbf{P}^{\mathbf{H_k}}_{\mathbf{c_k,A_k}}(\vec{\mathbf{u}}_\mathbf{k})$, $(\vec{d}_k, \beta_k)$, is obtained by solving (4.26)-(4.27).

Then there exists a neighbourhood $\mathcal{N}_1 \subset \mathcal{N}$ of $\bar{\vec{u}}$ and a finite $\bar{k}$ such that if $\vec{u}_k \in \mathcal{N}_1$ and $k \geq \bar{k}$ we also have $\beta_k = 0$.

Proof.

Assume that for infinitely many $k \in K$ we have $\beta_k \neq 0$. According to *Theorem 3.1* there exists $\bar{c} < \infty$ such that $c_k = \bar{c}$ for all $k \in K$. Furthermore, $(\vec{d}_k, \beta_k, \lambda_k)$ is the solution to the equations

$$\begin{aligned} \begin{bmatrix} H_k \vec{d}_k \\ 0 \end{bmatrix} + \tilde{A}_k^T \lambda_k + \begin{bmatrix} \nabla \bar{F}_0(\vec{u}_k) \\ 0 \end{bmatrix} &= \begin{bmatrix} 0 \\ -\bar{c} \end{bmatrix} \\ \tilde{A}_k \begin{bmatrix} d_k \\ \beta_k \end{bmatrix} &= \tilde{a}_k \end{aligned}$$

where $\tilde{A}_k$ is defined as $\tilde{A}_k$ in (4.9) with the exception that indices $E^+(\vec{u}_k)$, $E^-(\vec{u}_k)$, $I(\vec{u}_k)$ and $T(\vec{u}_k)$ are used. λ_k is the vector of the Lagrange multipliers.

First, we show that

$$\lim_{k \to \infty} \left\| \vec{d}_k \right\| = 0. \tag{4.28}$$

Indeed, from *Theorem 3.1*, we know that

$$\lim_{k \to \infty} \sigma^{H_k}_{c_k, A_k}(\vec{u}_k) = 0$$

and because we also have (Note that $\vec{d} = 0$ is feasible for $\mathbf{P}^{\mathbf{H}}_{\mathbf{c,A}}(\vec{\mathbf{u}})$.)

$$\sigma^{H_k}_{c_k, A_k}(\vec{u}_k) \leq -1/2(\vec{d}_k)^T H_k \vec{d}_k \leq 0,$$

from $(\mathbf{BH^N})$, we come to (4.28).

Since $\tilde{A}_k \tilde{A}_k^T$ is nonsingular in some neighbourhood $\mathcal{N}_0 \subset \mathcal{N}$ of $\bar{\vec{u}}$,

$$\lambda_k = -\left[\tilde{A}_k \tilde{A}_k^T\right]^{-1} \tilde{A}_k \begin{bmatrix} H_k \vec{d}_k + \nabla \bar{F}_0(\vec{u}_k) \\ \bar{c} \end{bmatrix},$$

(4.28) holds, H_k is bounded, $\tilde{A}_k$ and $\nabla \bar{F}_0$ are continuous, we can show that there exists a neighbourhood $\mathcal{N}_1 \subset \mathcal{N}_0$ of $\bar{\vec{u}}$ such that if $\vec{u}_k \in \mathcal{N}_1$ then λ_k is 'close' to $\bar{\lambda}$. Here $\bar{\lambda}$ satisfies necessary optimality conditions for problem (4.19)-(4.22):

$$\begin{bmatrix} \nabla \bar{F}_0(\bar{\vec{u}}) \\ 0 \end{bmatrix} + \bar{A}^T \bar{\lambda} = \begin{bmatrix} 0 \\ -\bar{c} \end{bmatrix}$$

(since (4.13) holds at $\vec{\bar{u}}$).

Let us specify in more detail the Lagrange multipliers for problem (4.19)–(4.22) at a stationary point $\vec{\bar{u}}$: $\bar{\lambda}_i^1$, $i \in E(\vec{\bar{u}})$, $\bar{\lambda}_j^2$, $j \in I(\vec{\bar{u}})$, $\bar{\lambda}(t)$, $t \in T(\vec{\bar{u}})$. We know that there exists k_1 such that for $k \geq k_1$

$$\begin{aligned} c_k = \bar{c} &= \sum_{i \in E^+(\vec{\bar{u}})} \bar{\lambda}_{i,1}^1 + \sum_{i \in E^-(\vec{\bar{u}})} \bar{\lambda}_{i,2}^1 + \sum_{j \in I(\vec{\bar{u}})} \bar{\lambda}_j^2 + \sum_{t \in T(\vec{\bar{u}})} \bar{\lambda}(t) \\ &> \sum_{i \in E(\vec{\bar{u}})} |\bar{\lambda}_i^1| + \sum_{i \in I(\vec{\bar{u}})} \bar{\lambda}_j^2 + \sum_{t \in T(\vec{\bar{u}})} \bar{\lambda}(t). \end{aligned} \quad (4.29)$$

Here, $\bar{\lambda}_{i,1}^1 \geq 0$, $i \in E^+(\vec{\bar{u}})$, $\bar{\lambda}_{i,2}^1 \geq 0$, $i \in E^-(\vec{\bar{u}})$, $\bar{\lambda}_j^2 \geq 0$, $j \in I(\vec{\bar{u}})$, $\bar{\lambda}(t) \geq 0$, $t \in T(\vec{\bar{u}})$. Furthermore, we have

$$\bar{\lambda}_{i,1}^1 - \bar{\lambda}_{i,2}^1 = \bar{\lambda}_i^1 \ \forall i \in E(\vec{\bar{u}}). \quad (4.30)$$

Arguments which lead to a strict inequality in (4.29) are quite subtle. Due to the fact that $\nabla\bar{h}_i^1(\vec{\bar{u}})$, $i \in E(\vec{\bar{u}})$, $\nabla\bar{h}_j^2(\vec{\bar{u}})$, $j \in I(\vec{\bar{u}})$, $\nabla\bar{q}(\vec{\bar{u}})(t)$, $t \in T(\vec{\bar{u}})$ are linearly independent, the Lagrange multipliers $\bar{\lambda}_i^1$, $i \in E(\vec{\bar{u}})$, $\bar{\lambda}_j^2$, $j \in I(\vec{\bar{u}})$, $\bar{\lambda}(t)$, $t \in T(\vec{\bar{u}})$ are uniquely defined and so is their sum on the right–hand side of (4.29). We remind that in order to obtain (4.29) one has to calculate $\{c_k\}$ according to (3.4). If c_k^{min} is defined as in *Remark 3.1* then any $c_k > c_k^{min}$ leads to the strict inequality of (4.29) because otherwise we would contradict the fact that the Lagrange multipliers are unique at the point $\vec{\bar{u}}$.

As we have shown above the Lagrange multipliers of problem (4.26)–(4.27) are arbitrarily close to the Lagrange multipliers of (4.19)–(4.22) at $\vec{\bar{u}}$. We know that $c_k = \bar{c}$ for $k \geq k_1$ and that together with (4.29) imply that there exists $\bar{k} \geq k_1$ such that for $k \geq \bar{k}$

$$c_k > \sum_{i \in E(\vec{\bar{u}})} |\tilde{\lambda}_i^{1,k}| + \sum_{i \in I(\vec{\bar{u}})} \tilde{\lambda}_j^{2,k} + \sum_{t \in T(\vec{\bar{u}})} \tilde{\lambda}^k(t), \quad (4.31)$$

where $\tilde{\lambda}_i^{1,k}$, $i \in E(\vec{\bar{u}})$, $\tilde{\lambda}_j^{2,k}$, $j \in I(\vec{\bar{u}})$, $\tilde{\lambda}^k(t)$, $t \in T(\vec{\bar{u}})$ are the Lagrange multipliers corresponding to direction finding subproblems whose linear constraints satisfy (4.10) (they are subproblems $\mathbf{P}_{\mathbf{A},\mathbf{H}}^{\mathbf{SQP}}(\vec{\mathbf{u}}_\mathbf{k})$ in which E, I and A are substitited by $E(\vec{\bar{u}})$, $I(\vec{\bar{u}})$ and $T(\vec{\bar{u}})$). However, because (**H5**) and (4.31) are satisfied, from Theorem 4.4.5 in [12], we know that $\beta_k = 0$. The lemma is proved. ∎

Lemma 4.1 has very important implications. It states that the scheme for updating penalty parameter c, based on the test function $t_{c,A}^H(u)$, guarantees that $\beta = 0$ in optimal solutions for the subproblems $\mathbf{P}_{\mathbf{c},\mathbf{A}}^{\mathbf{H}}(\vec{\mathbf{u}})$. This is exactly what we need to transform *Algorithm 4* into an SQP algorithm (c.f. [12]) . This implies that after a finite number of iterations subproblems $\mathbf{P}_{\mathbf{c}_\mathbf{k},\mathbf{A}_\mathbf{k}}^{\mathbf{H}_\mathbf{k}}(\vec{\mathbf{u}}_\mathbf{k})$ do not depend on the penalty parameter.

Another important feature of the update is the fact that c is not compared to the Lagrange multipliers of the subproblem $\mathbf{P}^{\mathbf{SQP}}_{\mathbf{A,H}}(\vec{\mathbf{u}})$. They can be difficult to calculate reliably when gradients of active constraints are 'almost' linearly dependent (c.f. [79]). We remind that the Lagrange multipliers of $\mathbf{P}^{\mathbf{H}}_{\mathbf{c,A}}(\vec{\mathbf{u}})$ are bounded between 0 and c.

On the other hand our scheme for updating c requires solving several $\mathbf{P}^{\mathbf{H}}_{\mathbf{c,A}}(\vec{\mathbf{u}})$ with different values of c (typically at first iterations of *Algorithm 4*). However, they can be solved efficiently at least if the QP method described in the Appendix is used.

The main result of this section refers to a superlinear rate of convergence defined as follows.

Definition 4.1 *The sequence $\{\vec{u}_k\}$ is superlinearly convergent to the point $\bar{\vec{u}}$ if there exists a sequence $\{\alpha_k\}$ converging to zero such that*

$$\left\|\vec{u}_{k+1} - \bar{\vec{u}}\right\| \leq \alpha_k \left\|\vec{u}_k - \bar{\vec{u}}\right\|, \quad k = 0, 1, \ldots \tag{4.32}$$

As noted in [36] one of the properties of superlinearly convergent sequences is

$$\lim_{k\to\infty} \frac{\|\vec{u}_{k+1} - \vec{u}_k\|}{\left\|\vec{u}_k - \bar{\vec{u}}\right\|} = 1 \tag{4.33}$$

provided $\vec{u}_k \neq \bar{\vec{u}} \ \ \forall k$.

Indeed, we have

$$\left|\|\vec{u}_{k+1} - \vec{u}_k\| - \left\|\vec{u}_k - \bar{\vec{u}}\right\|\right| \leq \left\|\vec{u}_{k+1} - \bar{\vec{u}}\right\|$$

which together with (4.32) lead to (4.33).

Theorem 4.1 *Let $\{\vec{u}_k\}$ be constructed by* Algorithm 4 *adapted for the problem* $(\mathbf{P_{NLP}}(\bar{\vec{\mathbf{u}}}))$ *and let the assumptions of* Lemma 4.1 *(excluding the assumption* (ii)*) be satisfied. Then*

(i) there exists a neighbourhood $\mathcal{N}$ of $\bar{\vec{u}}$ and a finite $\hat{k}$ such that if $\vec{u}_{\bar{k}} \in \mathcal{N}$ for some $\bar{k} \geq \hat{k}$ then $\vec{u}_k \in \mathcal{N}$ for all $k \geq \bar{k}$ and $\vec{u}_k \to_{k\to\infty} \bar{\vec{u}}$,

(ii) if, in addition, the modification (**M2**) *is applied in* Algorithm 4 *and*

$$\lim_{k\to\infty} \left[H_k - \bar{H}\right] \bar{Z} = 0, \tag{4.34}$$

where the matrix $\bar{Z}$ is defined by orthonormal vectors spanning the space

$$\begin{aligned} \{\vec{z} \in \mathcal{R}^{mN} : \langle \nabla \bar{h}^1_i(\bar{\vec{u}}), \vec{z}\rangle &= 0 \ \ \forall i \in E(\bar{\vec{u}}), \\ \langle \nabla \bar{h}^2_j(\bar{\vec{u}}), \vec{z}\rangle &= 0 \ \ \forall j \in I(\bar{\vec{u}}), \\ \langle \nabla \bar{q}(\bar{\vec{u}})(t), \vec{z}\rangle &= 0 \ \ \forall t \in T(\bar{\vec{u}})\}, \end{aligned}$$

then $\{\vec{u}_k\}$ is superlinearly convergent. Here, $\bar{H}$ is the Hessian of the Lagrangian as defined in (**H4**).

Proof.

$(\vec{d}_k, \lambda_k)$ is the solution of the following equations:

$$\nabla \bar{F}_0(\vec{u}_k) + \bar{A}_k^T \lambda_k + H_k \vec{d}_k = 0, \quad (4.35)$$

$$\bar{A}_k \vec{d}_k = \bar{a}_k, \quad (4.36)$$

where $\bar{A}_k$, $\bar{a}_k$ are defined as in (4.10) with the exception that indices $E(\bar{\vec{u}})$, $I(\bar{\vec{u}})$ and $T(\bar{\vec{u}})$ are used. This follows from the strict complementarity assumption (4.14). Furthermore, due to *Lemma 4.1*, there exists $\bar{k}$ such that for $k \geq \bar{k}$ $\beta_k = 0$ and we can solve the subproblems $\mathbf{P}^{\mathbf{SQP}}_{\mathbf{A,H}}(\vec{\mathbf{u}}_\mathbf{k})$ instead of $\mathbf{P}^{\mathbf{H}}_{\mathbf{c,A}}(\vec{\mathbf{u}}_\mathbf{k})$.

At the point $\bar{\vec{u}}$ we also have

$$\nabla \bar{F}_0(\bar{\vec{u}}) + \bar{A}^T \bar{\lambda} = 0, \quad (4.37)$$

$$\bar{a} = 0, \quad (4.38)$$

where $\bar{a}$ is the violation of the constraints at $\bar{\vec{u}}$ and $\bar{A}$ is evaluated at $\bar{\vec{u}}$. Equations (4.35)–(4.38) will be used to show that there exists $M < \infty$ such that

$$\left\| \vec{d}_k \right\| < M \left\| \vec{u}_k - \bar{\vec{u}} \right\|.$$

In order to prove that we recall Lemma 4 in [88], p. 14 (see also Lemma 4.1 in [30] and Lemma 5.1 in [17]). According to that lemma there exists $M_1 < \infty$ such that

$$\left\| \vec{d}_k \right\| \leq M_1 \left[\|\bar{a}_k\| + \left\| P_k \nabla \bar{F}_0(\vec{u}_k) \right\| \right], \quad (4.39)$$

where

$$P_k = I - \bar{A}_k \left[\bar{A}_k^T \bar{A}_k \right]^{-1} \bar{A}_k^T$$

is the projection matrix.

Since condition (4.37) can be restated as $\bar{P} \nabla \bar{F}_0(\bar{\vec{u}}) = 0$ (where $\bar{P}$ is the projection matrix evaluated at $\bar{\vec{u}}$) and (4.38) holds, (4.39) implies

$$\begin{aligned} \left\| \vec{d}_k \right\| &\leq M_1 \left[\|\bar{a}_k - \bar{a}\| + \left\| P_k \nabla \bar{F}_0(\vec{u}_k) - \bar{P} \nabla \hat{F}_0(\bar{\vec{u}}) \right\| \right] \\ &< M \left\| \vec{u}_k - \bar{\vec{u}} \right\|, \quad M < \infty \end{aligned}$$

since all functions are sufficiently smooth.

Define the open set which is a neighbourhood of $\bar{\vec{u}}$:

$$\mathcal{M} = \{ \vec{u} : \quad \left\| \vec{u} - \bar{\vec{u}} \right\| < \bar{\varepsilon}, \quad \bar{F}_c^R(\vec{u}) < \bar{F}_c^R(\bar{\vec{u}}) + \gamma_2 \left[\bar{\varepsilon}/(1 + sM) \right]^2 \}$$

(c.f. (4.23)). Here, $\varepsilon \geq \bar{\varepsilon} > 0$ (ε is as specified in (4.25)) and s is such that

$$\alpha_k \leq s \leq 1 \quad \forall k.$$

From (4.25), if $\vec{u} \in \mathcal{M}$, then we have

$$\gamma_2 \left\|\vec{u} - \bar{\vec{u}}\right\|^2 \leq \bar{F}_c^R(\vec{u}) - \bar{F}_c^R(\bar{\vec{u}}) \leq \gamma_2 \left[\bar{\varepsilon}/(1 + sM)\right]^2$$

thus

$$\left\|\vec{u} - \bar{\vec{u}}\right\| < \bar{\varepsilon}/(1 + sM). \tag{4.40}$$

On the other hand, we have

$$\begin{aligned} \left\|\vec{u}_{\bar{k}+1} - \bar{\vec{u}}\right\| &= \left\|\vec{u}_{\bar{k}} - \bar{\vec{u}} + \alpha_{\bar{k}}\vec{d}_{\bar{k}}\right\| \\ &\leq \left\|\vec{u}_{\bar{k}} - \bar{\vec{u}}\right\| + s\left\|\vec{d}_{\bar{k}}\right\|. \end{aligned}$$

Therefore,

$$\left\|\vec{u}_{\bar{k}+1} - \bar{\vec{u}}\right\| \leq (1 + sM)\left\|\vec{u}_{\bar{k}} - \bar{\vec{u}}\right\|$$

and if $\vec{u}_{\bar{k}} \in \mathcal{M}$, from (4.40), we obtain

$$\left\|\vec{u}_{\bar{k}+1} - \bar{\vec{u}}\right\| < \bar{\varepsilon}.$$

Furthermore,

$$\bar{F}_c^R(\vec{u}_{\bar{k}+1}) \leq \bar{F}_c^R(\vec{u}_{\bar{k}}) < \gamma_2 \left[\bar{\varepsilon}/(1 + sM)\right]^2$$

and that means that $\vec{u}_{\bar{k}+1} \in \mathcal{M}$, therefore $\vec{u}_k \in \mathcal{M} \;\; \forall k \geq \bar{k}$.

If we denote by $\bar{\mathcal{M}}$ a closure of the set $\mathcal{M}$ then $\{\vec{u}_k\}_{k \geq \bar{k}}$ has at least one limit point which lies in $\bar{\mathcal{M}}$. Because $\bar{F}_c^R(\cdot)$ is strongly convex on $\bar{\mathcal{M}}$ (c.f. (4.25), (4.29)) any limit point of $\{\vec{u}_k\}_{k \geq \bar{k}}$ must be equal to $\bar{\vec{u}}$.

We can easily show that the part *(i)* of the theorem's thesis holds also for the sequence $\{\vec{u}_k\}$ generated by *Algorithm 4* applied with the modification (**M2**). Since *Algorithm 4* (with the (**M2**) modification), when applied to problem (4.19)–(4.22), due to (4.34), has a superlinear rate of convergence (see, for example, Theorem 4.4.32 in [12]) we have proved the theorem. ■

Remark 4.1 *The conclusion* (i) *of the theorem remains valid for any SQP algorithm which satisfies the following conditions:*

1) it is locally convergent,

3) it has a direction finding subproblem (4.35)–(4.36),

3) H_k satisfy the conditions

a) $\|H_k^{-1}\| \leq M_1 < \infty$ *for all* k,

b) $\|H_k\| \leq M_2 < \infty$ *for all* k.

The condition a) can be replaced by

$$d^T H_k d \geq \nu \|d\|^2 \tag{4.41}$$

for all $d \neq 0$ *such that* $\bar{A}_k d = 0$. *Here,* $\nu > 0$.

Note that condition a) (or (4.41)) is required for Lemma 4 in [88] which is used in the proof of Theorem 4.1.

The conclusion (ii) *of the theorem is obviously valid if in addition an SQP algorithm is superlinearly convergent. The SQP algorithm presented in [74] uses the secant update for* H_k *and satisfies all of these requirements. This method is based on the finite differences update of the Hessian of the Lagrangian with the finite differences stepsize going to zero as* $k \to \infty$.

One could wonder why we fix the indices $T(\bar{\vec{u}})$, $E(\bar{\vec{u}})$, etc. which define constraints in the problem $(\mathbf{P_{NLP}}(\bar{\vec{\mathbf{u}}}))$. One could think that using only *a finite* number of constraints could be sufficient to guarantee, under assumptions (**H4**) and *(i)* and *(ii)* of *Theorem 4.1*, superlinear convergence of $\{\vec{u}_k\}$. Unfortunately this is not the case. The reason is that all known results on superlinear convergence of algorithms for constrainted problems (at least those known to the author) require that gradients $\nabla\bar{q}(\vec{u}_k)(t)$, $t \in T(\vec{u}_k)$, $\nabla\bar{h}_i^1(\vec{u}_k)$, $i \in E(\vec{u}_k)$, etc., change 'continuously' around local solution $\bar{\vec{u}}$ and that cannot be guaranteed if sets $T(\vec{u}_k)$, $E(\vec{u}_k)$, etc., are not fixed.

4.3 The Watchdog Technique for Redundant Constraints

The analysis of local properties of the problem $(\mathbf{P^N})$, with the help of the problem $(\mathbf{P_{NLP}}(\bar{\vec{\mathbf{u}}}))$, has been carried out under the assumption that $E(\bar{\vec{u}})$, $I(\bar{\vec{u}})$ and $T(\bar{\vec{u}})$ do not change in $(\mathbf{P_{NLP}}(\bar{\vec{\mathbf{u}}}))$. Using fix instances of t from $T(\bar{\vec{u}})$ is not adequate if the function of the state constraint $\bar{q}(\vec{u}_k)(\cdot)$ has local maximas defined as follows.

Definition 4.2 *The point* t^m *is a local maximum point of* $\bar{q}(\vec{u})(\cdot)$ *if there exists a neighbourhood of* t^m: $\mathcal{B}(t^m, \varepsilon) \subset T$, $\varepsilon > 0$, *such that* $\bar{q}(\vec{u})(t) < \bar{q}(\vec{u})(t^m)$ *for all* $t \in \mathcal{B}(t^m, \varepsilon) \setminus \{t^m\}$. *We denote by* $\mathcal{M}(\vec{u})$ *the set of all local maximum points of* $\bar{q}(\vec{u})(\cdot)$.

We assume that

(**H6**) For any local solution to the problem $(\mathbf{P^N})$, $\bar{\vec{u}}$, there exists a neighbourhood $\mathcal{B}(\bar{\vec{u}}, \varepsilon)$ of $\bar{\vec{u}}$, $\varepsilon > 0$, such that for any $t^m \in \mathcal{M}(\bar{\vec{u}})$ there exists a continuously differentiable function $m^t(\bar{\vec{u}}, \cdot) : \mathcal{B}(\bar{\vec{u}}, \varepsilon) \to T$ with the properties:

a) $m^t(\bar{\vec{u}}, \bar{\vec{u}}) = t^m$,

b) if $\mathcal{M}(\vec{\bar{u}}) = \{t_l^m : l \in \mathcal{K}(\vec{\bar{u}})\}$ then $\mathcal{M}(\vec{u}) = \{m_l^t(\vec{\bar{u}}, \vec{u}) : l \in \mathcal{K}(\vec{\bar{u}})\}$ —here $\vec{u} \in \mathcal{B}(\vec{\bar{u}}, \varepsilon)$ and $m_l^t(\vec{\bar{u}}, \cdot)$ corresponds to t_l^m,

c) function

$$\bar{q}_m(\vec{\bar{u}}, \cdot) := \bar{q}(\cdot)(m^t(\vec{\bar{u}}, \cdot))$$

is twice continuously differentiable on $\mathcal{B}(\vec{\bar{u}}, \varepsilon)$.

The conditions under which (**H6**) holds are discussed in [52] (pp. 644–645). The part b) states that $m_l^t(\vec{\bar{u}}, \vec{u})$, $l \in \mathcal{K}(\vec{\bar{u}})$ define all local maximum points for the function $\bar{q}(\vec{u})(\cdot)$ in some neighbourhood of $\vec{\bar{u}}$.

It is important to note that under these conditions (which we do not introduce here in order not to further complicate the presentation) we also have

$$\nabla \bar{q}_m^l(\vec{\bar{u}}, \vec{u}) = \nabla \bar{q}(\vec{u})(m_l^t(\vec{\bar{u}}, \vec{u}))$$

where

$$\bar{q}_m^l(\vec{\bar{u}}, \vec{u}) := \bar{q}(\vec{u})(m_l^t(\vec{\bar{u}}, \vec{u}))$$

and $\nabla \bar{q}_m^l(\vec{\bar{u}}, \vec{u})$ is the gradient of $\bar{q}_m^l$ with respect to $\vec{u}$. This is important from the computational point of view. (We do not need to know the function $m^t(\vec{\bar{u}}, \cdot)$ and its gradient in order to evaluate the gradient of $\bar{q}_m(\vec{\bar{u}}, \cdot)$.)

If functions $\bar{q}(\vec{u})(\cdot)$ have local maximum points, constraints (4.21) need to be modified by substituting the set $T(\vec{\bar{u}})$ by two sets: $T^c(\vec{\bar{u}})$ and

$$T^v(\vec{\bar{u}}) := \{t_l^m : l \in \mathcal{K}^s(\vec{\bar{u}}) \subset \mathcal{K}(\vec{\bar{u}})\}$$

(not all local maximum points of $\bar{q}(\vec{\bar{u}})(\cdot)$ have to be included in $T(\vec{\bar{u}})$). Furthermore, we request

$$\begin{aligned} T^c(\vec{\bar{u}}) \bigcup T^v(\vec{\bar{u}}) &= T(\vec{\bar{u}}) \\ T^c(\vec{\bar{u}}) \bigcap T^v(\vec{\bar{u}}) &= \emptyset \\ T^c(\vec{\bar{u}}) \bigcap \mathcal{M}(\vec{\bar{u}}) &= \emptyset. \end{aligned}$$

Next, the problem ($\mathbf{P_{NLP}}(\vec{\bar{\mathbf{u}}})$) is adapted to the problem in which constraints (4.21) are substituted by:

$$\begin{aligned} \bar{q}(\vec{u})(t) &\leq 0 \quad \forall t \in T^c(\vec{\bar{u}}) \\ \bar{q}(\vec{u})(m_l^t(\vec{\bar{u}}, \vec{u})) &\leq 0 \quad \forall l \in \mathcal{K}^s(\vec{\bar{u}}). \end{aligned}$$

Notice that the thesis of *Theorem 4.1* holds also for the problem ($\mathbf{P_{NLP}}(\vec{\bar{\mathbf{u}}})$) modified in this way. This follows from the fact that close to $\vec{\bar{u}}$, $\nabla \bar{q}_m^l(\vec{\bar{u}}, \vec{u}_k)$, $l \in \mathcal{K}^s(\vec{\bar{u}})$ are continuously differentiable functions with respect to $\vec{u}_k$.

Functions m^t can also be defined for any $\vec{v}$ in some neighbourhood of a local solution to the problem $(\mathbf{P^N})$. We can introduce $T^c(\vec{v})$, $T^v(\vec{v})$ and the corresponding set of indices $\mathcal{K}^s(\vec{v}) \subset \mathcal{K}(\vec{v})$ such that $T^v(\vec{v}) = \{t_l^m : l \in \mathcal{K}^s(\vec{v})\}$ and the corresponding set of functions $m_l^t(\vec{v}, \cdot), l \in \mathcal{K}^s(\vec{v})$ define the constraints:

$$\bar{q}(\vec{u})(m_l^t(\vec{v}, \vec{u})) \leq 0 \quad \forall l \in \mathcal{K}^s(\vec{v}). \tag{4.42}$$

Furthermore, the following relations hold:

$$\begin{aligned} T^c(\vec{v}) \bigcup T^v(\vec{v}) &= T(\vec{v}) \\ T^c(\vec{v}) \bigcap T^v(\vec{v}) &= \emptyset \\ T^c(\vec{v}) \bigcap \mathcal{M}(\vec{v}) &= \emptyset. \end{aligned}$$

Constraints (4.42) are used in the algorithm stated below.

Theorem 4.1 can be used to construct a superlinearly convergent algorithm for the problem $(\mathbf{P^N})$. The algorithm switches to a method which uses a fixed number of constraints once close enough neighbourhood of a stationary point to the problem $(\mathbf{P^N})$ has been identified. If the stationary point is also a local solution to the problem and H_k 'appropriately' approaches the matrix $\bar{H}$ as defined in **(H4)** then according to *Theorem 4.1* we can expect that the algorithm will converge superlinearly to the solution of the problem $(\mathbf{P^N})$.

FD Algorithm

Parameters: $\{L_k^{h^1}(i)\}_{k=1}^{\infty} \,:\, 0 < L_k^{h^1}(i) < \infty \ \forall k, \ \forall i \in E, \ \{L_k^{h^2}(j)\}_{k=1}^{\infty} \,:\, 0 < L_k^{h^2}(j) < \infty \ \forall k, \ \forall j \in I, \ \{L_k^q(t)\}_{k=1}^{\infty} \,:\, 0 < L_k^q(t) < \infty \ \forall k, \ \forall t \in T$, $0 < \psi_1, \psi_2 \ll 1, \ 1 < \varrho, \ \varepsilon > 0, \ H_k, \ \ k = 0, 1, \ldots$

1. Choose the initial control $\vec{u}_0 \in \mathcal{U}$, calculate $M(\vec{u}_0)$, set $\sigma_{c_{-1}, A_{-1}}^{H_{-1}}(\vec{u}_{-1}) = -\infty$, $k = 0$. Set the formulation of the problem **(fop)** to *standard*.

2. If $M(\vec{u}_k) \leq \psi_1$ and $\sigma_{c_{k-1}, A_{k-1}}^{H_{k-1}}(\vec{u}_{k-1}) \geq -\psi_2$ set **fop** to *reduced*.

3. If **fop** is *standard* find the direction $\vec{d}_k$, the estimate of the directional derivative, $\sigma_{c_k, A_k}^{H_k}(\vec{u}_k)$, and the next control $\vec{u}_{k+1}$ as in *Algorithm 4*. Substitute $\vec{u}_k$ for $\vec{v}$, $E^+(\vec{u}_k)$, for E_r^+, $E^-(\vec{u}_k)$ for E_r^-, $I(\vec{u}_k)$ for I_r, $T^c(\vec{u}_k)$ for T_r^c, $\mathcal{K}^s(\vec{u}_k)$ for $\mathcal{K}_r^s$ and k for k_s.

4. If **fop** is *reduced* find $\vec{d}_k$, $\sigma_{c_k, A_k}^{H_k}(\vec{u}_k)$ and the next control $\vec{u}_{k+1}$ by using only the constraints $\bar{h}_{i,1}^1(\vec{u}_k)$, $i \in E_r^+$, $\bar{h}_{i,2}^1(\vec{u}_k)$, $i \in E_r^-$, $\bar{h}_j^2(\vec{u}_k)$, $j \in I_r$, $\bar{q}(\vec{u}_k)(t)$, $t \in T_r^c$ and $\bar{q}(\vec{u}_k)(m_l^t(\vec{v}, \vec{u}_k))$, $l \in \mathcal{K}_r^s$.

5. If **fop** is *reduced* and

$$\begin{aligned}\bar{q}(\vec{u}_k)(t) &> \varrho L_k^q(t) \text{ for some } t \in R_{\epsilon,\vec{u}_k}, \text{ or}\\ \bar{h}_i^1(\vec{u}_k) &> \varrho L_k^{h^1}(i) \text{ for some } i \in E, \text{ or}\\ \bar{h}_j^2(\vec{u}_k) &> \varrho L_k^{h^2}(j) \text{ for some } j \in I,\end{aligned}$$

set **fop** to *standard*, substitute k_s for k, $\vec{v}$ for $\vec{u}_k$ and go to Step 3.

6. Increase k by one and go to Step 2.

FD Algorithm applies a kind of the watchdog technique (c.f. [31]). If using only a fixed number of constraints does not lead to an excessive violation of constraints which have not been included in the problem, we continue solving the problem. If the constraints are violated by some margin we return to solving the original problem but from the latest control which has been used in the *standard* mode. This means all iterations between the current iteration and the latest *standard* iteration are 'lost'. Therefore, we should use $L_k^q(t)$, $L_k^{h^1}(i)$, $L_k^{h^2}(j)$ which give as small values as possible. Notice that

$$L_k^q(t) = L\|\vec{d}_k\|,$$

where L is such that $\|\nabla\bar{q}(\vec{u}_k)(t)\| \leq L$, would be worse choice for the monitoring function than

$$L_k^q(t) = |\langle\nabla\bar{q}(\vec{u}_k)(t), \vec{d}_k\rangle|.$$

The following lemma is crucial in constructing adequate functions $L_k(\cdot)$. The lemma concerns a relative rate of convergence of sequences $\{\langle\vec{z}_k, \vec{u}_k - \bar{\vec{u}}\rangle\}$, $\{\langle\vec{z}_k, \vec{d}_k\rangle\}$ for some convergent sequence $\{\vec{z}_k\}$ and mimics the well-known result (4.33) when $\{\vec{u}_k\}$ is superlinearly convergent to $\bar{\vec{u}}$ and $\vec{u}_{k+1} = \vec{u}_k + \vec{d}_k$.

Before presenting the lemma we introduce notations which are used in its statement. We denote by λ_k the angle between $\vec{u}_k - \bar{\vec{u}}$ and $\vec{u}_k - \vec{u}_{k+1}$, ϕ_k the angle between $\vec{u}_k - \bar{\vec{u}}$ and $\vec{z}_k$ and γ_k the angle between $\vec{d}_k$ and $\vec{z}_k$.

Lemma 4.2 *If the sequence* $\vec{u}_{k+1} = \vec{u}_k + \vec{d}_k$ *is superlinearly convergent to* $\bar{\vec{u}}$, $\{\vec{z}_k\}$ *is a convergent sequence and the following holds:*

$$\vec{d}_k \neq 0, \quad \vec{u}_k - \bar{\vec{u}} \neq 0 \ \ \forall k,$$

then

$$\lim_{k\to\infty} \phi_k = \lim_{k\to\infty} \gamma_k, \tag{4.43}$$

and, if in addition,

$$\left\langle \vec{z}_k, \vec{d}_k \right\rangle \neq 0, \ \ \langle \vec{z}_k, \vec{u}_k - \bar{\vec{u}}\rangle \neq 0 \ \ \forall k,$$

$$\lim_{k\to\infty} \frac{\|\vec{u}_{k+1} - \vec{\bar{u}}\|}{\langle \vec{z}_k, \vec{u}_k - \vec{\bar{u}} \rangle} = 0, \tag{4.44}$$

we also have

$$\lim_{k\to\infty} \frac{\left|\left\langle \vec{z}_k, \vec{d}_k \right\rangle\right|}{\left|\langle \vec{z}_k, \vec{u}_k - \vec{\bar{u}} \rangle\right|} = 1. \tag{4.45}$$

Proof.

Because

$$\begin{aligned} \gamma_k &\le \phi_k + \lambda_k \\ \phi_k &\le \gamma_k + \lambda_k \end{aligned}$$

(c.f. Figure 5.1), in order to prove (4.43) it is sufficient to show that

$$\lim_{k\to\infty} \cos(\lambda_k) = 1. \tag{4.46}$$

Then

$$\lim_{k\to\infty} \gamma_k \le \lim_{k\to\infty} \phi_k, \qquad \lim_{k\to\infty} \gamma_k \ge \lim_{k\to\infty} \phi_k$$

and

$$\lim_{k\to\infty} \gamma_k = \lim_{k\to\infty} \phi_k,$$

as required.

To prove (4.46) notice that

$$\begin{aligned} \lim_{k\to\infty} \cos(\lambda_k) &= \lim_{k\to\infty} \frac{\langle \vec{u}_k - \vec{\bar{u}}, \vec{u}_k - \vec{u}_{k+1} \rangle}{\|\vec{u}_k - \vec{\bar{u}}\| \left\|\vec{d}_k\right\|} \\ &= \lim_{k\to\infty} \frac{\langle \vec{u}_{k+1} - \vec{\bar{u}}, \vec{u}_k - \vec{u}_{k+1} \rangle}{\|\vec{u}_k - \vec{\bar{u}}\| \left\|\vec{d}_k\right\|} + \\ &\quad \lim_{k\to\infty} \frac{\left\langle \vec{d}_k, \vec{d}_k \right\rangle}{\|\vec{u}_k - \vec{\bar{u}}\| \left\|\vec{d}_k\right\|} \\ &= \lim_{k\to\infty} \frac{\langle \vec{u}_{k+1} - \vec{\bar{u}}, \vec{u}_k - \vec{u}_{k+1} \rangle}{\|\vec{u}_k - \vec{\bar{u}}\| \left\|\vec{d}_k\right\|} + 1. \end{aligned} \tag{4.47}$$

The last equality follows from

$$\lim_{k\to\infty} \frac{\left\|\vec{d}_k\right\|}{\|\vec{u}_k - \vec{\bar{u}}\|} = 1 \tag{4.48}$$

which is the restatement of (4.33).

Furthermore, we have

$$\lim_{k\to\infty} \frac{\left|\langle \vec{u}_k - \vec{u}_{k+1}, \vec{u}_{k+1} - \vec{\bar{u}}\rangle\right|}{\left\|\vec{d}_k\right\| \left\|\vec{u}_k - \vec{\bar{u}}\right\|} \leq \lim_{k\to\infty} \frac{\left\|\vec{u}_{k+1} - \vec{\bar{u}}\right\|}{\left\|\vec{u}_k - \vec{\bar{u}}\right\|} = 0$$

and that together with (4.47) prove (4.46).

In order to prove (4.45) we consider the limit

$$\lim_{k\to\infty} \frac{|\cos(\gamma_k)|}{|\cos(\phi_k)|} = \lim_{k\to\infty} \frac{\left\|\vec{x}_k^1\right\| \left\|\vec{u}_k - \vec{\bar{u}}\right\|}{\left\|\vec{x}_k^2\right\| \left\|\vec{d}_k\right\|},$$

where $\vec{x}_k^1 = P_{z_k}[-\vec{d}_k]$, $\vec{x}_k^2 = P_{z_k}[\vec{u}_k - \vec{\bar{u}}]$ and P_{z_k} is the orthogonal projection on the subspace spanned by z_k.

The next step in the proof is to show that

$$\lim_{k\to\infty} \frac{\left\|\vec{x}_k^1\right\| \left\|\vec{u}_k - \vec{\bar{u}}\right\|}{\left\|\vec{x}_k^2\right\| \left\|\vec{d}_k\right\|} = 1. \tag{4.49}$$

This follows from (4.48) and

$$\left|\left\|\vec{x}_k^1\right\| - \left\|\vec{x}_k^2\right\|\right| \leq \left\|P_{z_k}\left[-\vec{d}_k + \vec{\bar{u}} - \vec{u}_k\right]\right\| \leq \left\|\vec{u}_{k+1} - \vec{\bar{u}}\right\|$$

which implies

$$\lim_{k\to\infty} \frac{\left\|\vec{x}_k^1\right\|}{\left\|\vec{x}_k^2\right\|} = 1$$

since

$$\lim_{k\to\infty} \frac{\left|\left\|\vec{x}_k^1\right\| - \left\|\vec{x}_k^2\right\|\right|}{\left\|\vec{x}_k^2\right\|} \leq \lim_{k\to\infty} \frac{\left\|\vec{z}_k\right\| \left\|\vec{u}_{k+1} - \vec{\bar{u}}\right\|}{\left|\langle \vec{z}_k, \vec{u}_k - \vec{\bar{u}}\rangle\right|} = 0$$

according to assumption (4.44).

Eventually

$$\lim_{k\to\infty} \frac{\left|\left\langle \vec{z}_k, \vec{d}_k\right\rangle\right|}{\left|\langle \vec{z}_k, \vec{u}_k - \vec{\bar{u}}\rangle\right|} = 1,$$

because

$$\begin{aligned}
\lim_{k\to\infty} \cos(\phi_k) &= \lim_{k\to\infty} \frac{\langle \vec{z}_k, \vec{u}_k - \vec{\bar{u}}\rangle}{\left\|\vec{z}_k\right\| \left\|\vec{u}_k - \vec{\bar{u}}\right\|}, \\
\lim_{k\to\infty} \cos(\gamma_k) &= \lim_{k\to\infty} \frac{\left\langle \vec{z}_k, \vec{d}_k\right\rangle}{\left\|\vec{z}_k\right\| \left\|\vec{d}_k\right\|}, \\
\lim_{k\to\infty} \frac{|\cos(\gamma_k)|}{|\cos(\phi_k)|} &= 1 \qquad \text{(due to (4.49))}
\end{aligned}$$

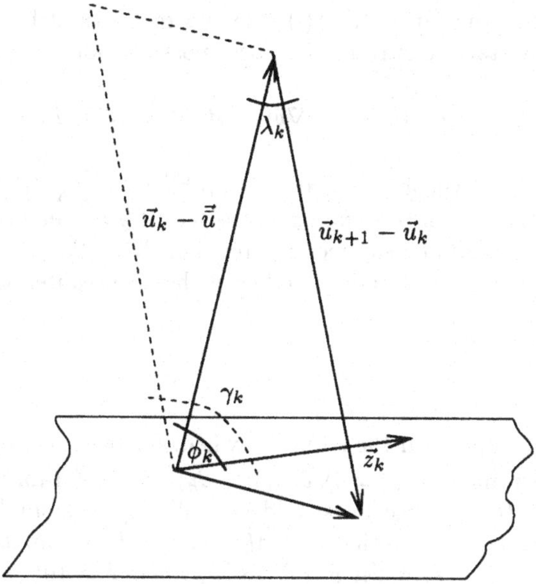

FIGURE 5.1. Geometry of a superlinearly convergent sequence.

and (4.48) holds.

■

In fact only the first part of the lemma's thesis, (4.43), is needed to justify the watchdog technique of *FD Algorithm*. The property of superlinearly convergent sequences expressed by (4.45) is provided to show that rephrasing (4.33), by using scalar products, is not allowed in general.

Define $L_k^q(\cdot)$, $L_k^{h^1}(\cdot)$, $L_k^{h^2}(\cdot)$ as follows

$$\begin{aligned}
L_k^q(t) &= \begin{cases} |\langle \nabla \bar{q}(\vec{u}_k)(t), \vec{d}_k \rangle| & \text{if } \dfrac{|\langle \nabla \bar{q}(\vec{u}_k)(t), \vec{d}_k \rangle|}{||\vec{d}_k||} > \nu > 0 \\ \nu ||\vec{d}_k|| & \text{if } \dfrac{|\langle \nabla \bar{q}(\vec{u}_k)(t), \vec{d}_k \rangle|}{||\vec{d}_k||} \le \nu \end{cases} \\
L_k^{h^1}(i) &= \begin{cases} |\langle \nabla \bar{h}_i^1(\vec{u}_k), \vec{d}_k \rangle| & \text{if } \dfrac{|\langle \nabla \bar{h}_i^1(\vec{u}_k), \vec{d}_k \rangle|}{||\vec{d}_k||} > \nu \\ \nu ||\vec{d}_k|| & \text{if } \dfrac{|\langle \nabla \bar{h}_i^1(\vec{u}_k), \vec{d}_k \rangle|}{||\vec{d}_k||} \le \nu \end{cases} \\
L_k^{h^2}(j) &= \begin{cases} |\langle \nabla \bar{h}_j^2(\vec{u}_k), \vec{d}_k \rangle| & \text{if } \dfrac{|\langle \nabla \bar{h}_j^2(\vec{u}_k), \vec{d}_k \rangle|}{||\vec{d}_k||} > \nu \\ \nu ||\vec{d}_k|| & \text{if } \dfrac{|\langle \nabla \bar{h}_j^2(\vec{u}_k), \vec{d}_k \rangle|}{||\vec{d}_k||} \le \nu . \end{cases}
\end{aligned} \tag{4.50}$$

Notice that the function $L_k^q(\cdot)$ provides really tight bounds for the violation of the state constraint. The best possible bounds are

$$\bar{q}(\vec{u}_k)(t) \leq -\left\langle \nabla\bar{q}(\vec{u}_k)(t), \vec{d}_k \right\rangle \quad \forall t \in R_{\varepsilon,\vec{u}_k}$$

and this happens when $\beta_k = 0$ and the problems $\mathbf{P}^{\mathbf{H_k}}_{\mathbf{c_k},\mathbf{A_k}}(\vec{\mathbf{u}}_\mathbf{k})$ with (possibly) an infinite number of constraints are solved. If $\varrho = 1$ in *Algorithm FD* then these bounds would be enforced in Step 5 of *Algorithm FD*. $\varrho > 1$ is needed for performing the nontrivial number of the *reduced* iterations.

The case

$$L_k^{\cdot}(\cdot) \leq \nu \left\|\vec{d}_k\right\|$$

is needed to cope with the situation when, for instance, $\langle \bar{q}(\vec{u}_k)(t), \vec{d}_k\rangle = 0$ which would make $L_k^q(t) = \langle \nabla\bar{q}(\vec{u}_k)(t), \vec{d}_k\rangle$ not useful for monitoring state constraints. In this case we should set value of ν as small as possible. If, for instance, we expect that *FD Algorithm* will terminate with $|(\vec{d}_k)_i| \leq \varepsilon_{STOP}$ $\forall i = 1, \ldots, mN$, set $\nu = \delta\varepsilon_{STOP}$, where $\delta \geq 10^1$, to guarantee that $\bar{q}(\vec{u}_k)(t)$ are second order terms w.r.t. $\|\vec{d}_k\|$ as the relation $\langle \nabla\bar{q}(\vec{u}_k)(t), \vec{d}_k\rangle = 0$ suggests.

We can prove the following theorem.

Theorem 4.2 *If the assumptions of* Theorem 4.1 *are satisfied, the assumption* (**H4**) *holds at every stationary point of the problem* ($\mathbf{P^N}$), *and*

(*i*) *for any stationary point of* ($\mathbf{P^N}$), $\bar{\vec{u}}$, *there exists a neighbourhood* $\mathcal{B}(\bar{\vec{u}}, \varepsilon)$, $\varepsilon > 0$, *such that for* $\vec{u}_k \in \mathcal{B}(\bar{\vec{u}}, \varepsilon)$ *and any set* $E^+(\vec{u}_k)$, $E^-(\vec{u}_k)$, $I(\vec{u}_k)$, $T(\vec{u}) = T^c(\vec{u}_k) \bigcup T^v(\vec{u}_k)$, $T^v(\vec{u}_k) = \{m_l^t(\bar{\vec{u}}, \vec{u}_k) : l \in \mathcal{K}^s(\bar{\vec{u}})\}$, *vectors* $\nabla\bar{h}_i^1(\bar{\vec{u}})$, $i \in E^+(\vec{u}_k) \bigcup E^-(\vec{u}_k)$, $\nabla\bar{h}_j^2(\bar{\vec{u}})$, $j \in I(\vec{u}_k)$, $\nabla\bar{q}(\bar{\vec{u}})(t)$, $t \in T^c(\vec{u}_k)$, $\nabla\bar{q}(\bar{\vec{u}})(m_l^t(\bar{\vec{u}}, \bar{\vec{u}}))$, $l \in \mathcal{K}^s(\bar{\vec{u}})$ *are linearly independent,*

(*ii*) $L_k^q(\cdot)$, $L_k^{h^1}(\cdot)$, $L_k^{h^2}(\cdot)$ *are defined by (4.50)*

then

1) *after a finite number of iterations a fixed number of constraints is used to define the sequence* $\{\vec{u}_k\}$,

2) $\{\vec{u}_k\}$ *is superlinearly convergent to a stationary point of the problem* ($\mathbf{P^N}$).

The assumption (*i*) is needed to exclude the case of constraints whose gradients are linearly independent in a neighbourhood of the stationary point $\bar{\vec{u}}$ but become linearly dependent at $\bar{\vec{u}}$.

Notice that functions m_l^t used in (*i*) are defined at $\bar{\vec{u}}$. That is justified by part b) of the assumption (**H6**).

The proof of the theorem is preceded by the lemma which characterizes consequences of the assumption *(i)* on the optimality conditions of the problem ($\mathbf{P^N}$). We have to rely on the lemma, in the proof of the theorem, since the assumption (**H4**) is applicable to *some* (not *any*) set of indices: $E(\vec{\bar{u}})$, $I(\vec{\bar{u}})$, $T(\vec{\bar{u}})$. Note that the set T_r^c used in the *reduced* mode of *FD Algorithm* does not have to be a part of $T(\vec{\bar{u}})$ which appears in (4.13). However, as we show in the lemma below, condition (4.13) will also be satisfied by T_r^c 'close' to that $T(\vec{\bar{u}})$.

We refer, in the lemma, to the set $\hat{T}(\vec{\bar{u}})$ which is ε–*close* to $T(\vec{\bar{u}})$ ($\varepsilon > 0$). The set $\hat{T}(\vec{\bar{u}})$ is ε–*close* to $T(\vec{\bar{u}})$ if there exists a one–to–one function $r : T(\vec{\bar{u}}) \to \hat{T}(\vec{\bar{u}}) \subset R_{0,\vec{\bar{u}}}$ such that

(i) $r(t) = t \quad \forall t \in T^v(\vec{\bar{u}})$,

(ii) $|r(t) - t| \leq \varepsilon \quad \forall t \in T(\vec{\bar{u}})$,

Lemma 4.3 *Assume that $\vec{\bar{u}}$ is a stationary point of* ($\mathbf{P^N}$) *at which the assumption* (**H4**) *and the assumption* (i) *of* Theorem 4.1 *together with* ($\mathbf{CQ^N}$) *are satisfied. Then there exists $\varepsilon > 0$ such that the set of vectors defined by $E(\vec{\bar{u}})$, $I(\vec{\bar{u}})$ and any $\hat{T}(\vec{\bar{u}})$ which is ε*–close *to $T(\vec{\bar{u}})$, the set of indices as specified in* (**H4**)*, satisfies:*

a) it is the maximum set of linearly independent vectors in some neighbourhood of $\vec{\bar{u}}$,

b) it meets the conditions specified in (**H4**).

Proof.

a) follows from the continuity of vectors $\nabla\bar{q}(\vec{\bar{u}})(\cdot)$.

b) Consider the problem:

$$\min_{\vec{d}} \left[\left\langle \nabla \bar{F}_0(\vec{\bar{u}}), \vec{d}\right\rangle + 1/2 \left\|\vec{d}\right\|^2\right] \tag{4.51}$$

$$\text{s. t. } \bar{A}\vec{d} = 0, \tag{4.52}$$

where $\bar{A}$ is defined by vectors $\nabla \bar{h}_i^1(\vec{\bar{u}})$, $i \in E(\vec{\bar{u}})$, $\nabla \bar{h}_j^2(\vec{\bar{u}})$, $j \in I(\vec{\bar{u}})$, $\nabla\bar{q}(\vec{\bar{u}})(t)$, $t \in \hat{T}(\vec{\bar{u}})$. $\hat{T}(\vec{\bar{u}})$ is ε–*close* to $T(\vec{\bar{u}})$.

If $\vec{\hat{d}}$ is the solution to problem (4.51)–(4.52) and is such that $\vec{\hat{d}} = 0$, then it is easy to show that (4.13) is satisfied for $E(\vec{\bar{u}})$, $I(\vec{\bar{u}})$ and $\hat{T}(\vec{\bar{u}})$ which is ε–close to $T(\vec{\bar{u}})$ (with ε sufficiently small). Furthermore, $\hat{T}(\vec{\bar{u}})$, $E(\vec{\bar{u}})$ and $I(\vec{\bar{u}})$ satisfy a). Otherwise we have $\vec{\hat{d}} \neq 0$ and the following holds:

$$\vec{\hat{d}} + \bar{A}^T\bar{\lambda} + \nabla\bar{F}_0(\vec{\bar{u}}) = 0, \tag{4.53}$$

$$\bar{A}\vec{\hat{d}} = 0. \tag{4.54}$$

According to a) (4.54) is also satisfied for any $\nabla\bar{q}(\vec{\bar{u}})(t)$, $t \in R_{0,\vec{\bar{u}}}$, $\nabla\bar{h}_i^1(\vec{\bar{u}})$, $i \in E$, $\nabla\bar{h}_j^2(\vec{\bar{u}})$, $j \in I_{0,\vec{\bar{u}}}$:

$$\left\langle \nabla\bar{q}(\vec{\bar{u}}), \vec{\bar{d}} \right\rangle = 0 \;\; \forall t \in R_{0,\vec{\bar{u}}} \quad (4.55)$$

$$\left\langle \nabla\bar{h}_i^1(\vec{\bar{u}}), \vec{\bar{d}} \right\rangle = 0 \;\; \forall i \in E \quad (4.56)$$

$$\left\langle \nabla\bar{h}_j^2(\vec{\bar{u}}), \vec{\bar{d}} \right\rangle = 0 \;\; \forall j \in I_{0,\vec{\bar{u}}}. \quad (4.57)$$

These equalities together with (4.53) contradict the fact that $\vec{\bar{u}}$ is a stationary point to the problem ($\mathbf{P}^{\mathbf{N}}$). (We could show that $\vec{\bar{d}} \neq 0$ is a direction of descent for the penalty function $\bar{F}_c$, note also that ($\mathbf{CQ}^{\mathbf{N}}$) holds.) Thus (i) of (**H4**) is satisfied if $\hat{T}(\vec{\bar{u}})$ is ε*–close* to $T(\vec{\bar{u}})$ and ε is sufficiently small.

Furthermore, the condition (ii) of (**H4**) is met because a) holds and second order derivatives of $\bar{q}(\vec{u})(t)$ are continuous with respect to $(\vec{u}, t)$. (When $\hat{T}(\vec{\bar{u}})$ is ε–close to $T(\vec{\bar{u}})$ then the corresponding Lagrange multipliers are close to each other. Note also (4.55)–(4.57).) ■

Proof.

(*Theorem 4.2*) The proof consists of three parts. In *Part 1)* we analyse the consequences of the fact that $\{\vec{u}_k\}$ is superlinearly convergent. In *Part 2)* we show that if $\vec{u}_{\bar{k}}$ is in some neighbourhood of a stationary point of the problem ($\mathbf{P}^{\mathbf{N}}$) at which the assumption (**H4**) is satisfied then $\{\vec{u}_k\}_{k \geq \bar{k}}$ is superlinearly convergent to $\vec{\bar{u}}$. Finally in *Part 3)* we prove that there exists a finite $\bar{k}$ such that $\vec{u}_{\bar{k}}$ falls into the neighbourhood specified in *Part 2)*.

Part 1). Assume that $\{\vec{u}_k\}$ is superlinearly convergent to $\vec{\bar{u}}$. We have

$$\begin{aligned} \bar{q}(\vec{\bar{u}})(t) = 0 \;\; = \;\; & \bar{q}(\vec{u}_k)(t) + \\ & \left\langle \nabla\bar{q}(\vec{u}_k)(t), \vec{\bar{u}} - \vec{u}_k \right\rangle + o(t, \vec{u}_k - \vec{\bar{u}}) \\ & \forall t \in R_{0,\vec{\bar{u}}} \end{aligned} \quad (4.58)$$

and similar relations for $\bar{h}_i^1(\vec{\bar{u}})$, $\bar{h}_j^2(\vec{\bar{u}})$.

Furthermore, $o(t, \vec{u}_k - \vec{\bar{u}})$ can be written as

$$o(t, \vec{u}_k - \vec{\bar{u}}) = \frac{1}{2} \left[\vec{u}_k - \vec{\bar{u}}\right]^T \nabla^2_{uu}\bar{q}(\vec{\xi}_k^1)(t) \left[\vec{u}_k - \vec{\bar{u}}\right], \quad \vec{\xi}_k^1 \in \left[\vec{u}_k, \vec{\bar{u}}\right],$$

where $[a, b]$ is a line segment spanned by vectors a, b.

From the Mean Value Theorem, we also have

$$\nabla\bar{q}(\vec{u}_k)(t) - \nabla\bar{q}(\vec{\bar{u}})(t) = \nabla^2_{uu}\bar{q}(\vec{\xi}_k^2)(t) \left[\vec{u}_k - \vec{\bar{u}}\right], \quad \vec{\xi}_k^2 \in \left[\vec{u}_k, \vec{\bar{u}}\right].$$

Therefore,

$$\begin{aligned} M_k &:= \langle \vec{u}_k - \vec{\bar{u}}, \nabla \bar{q}(\vec{u}_k)(t) - \nabla \bar{q}(\vec{\bar{u}})(t) \rangle \\ &= \left[\vec{u}_k - \vec{\bar{u}}\right]^T \nabla^2_{uu}\bar{q}(\vec{\xi}^1_k)(t) \left[\vec{u}_k - \vec{\bar{u}}\right] + \\ &\quad \left[\vec{u}_k - \vec{\bar{u}}\right]^T F_k(t) \left[\vec{u}_k - \vec{\bar{u}}\right]. \end{aligned} \tag{4.59}$$

Here,

$$F_k(t) = \nabla^2_{uu}\bar{q}(\vec{\xi}^2_k)(t) - \nabla^2_{uu}\bar{q}(\vec{\xi}^1_k)(t) \text{ and } \lim_{k\to\infty} F_k(t) = 0, \tag{4.60}$$

since

$$\lim_{k\to\infty} \vec{\xi}^1_k = \lim_{k\to\infty} \vec{\xi}^2_k = \vec{\bar{u}}$$

and $\nabla^2_{uu}\bar{q}(\cdot)(t)$ is continuous according to (**H4**).

If for infinitely many k

$$\frac{\left\langle \nabla \bar{q}(\vec{u}_k)(t), \vec{d}_k \right\rangle}{\left\| \vec{d}_k \right\|} > \nu \tag{4.61}$$

($\nu > 0$ as in (4.50)) then, from (4.43) (assuming $\vec{z}_k := \nabla \bar{q}(\vec{u}_k)(t)$), we can show that

$$\lim_{k\to\infty} \frac{\langle \nabla \bar{q}(\vec{u}_k)(t), \vec{u}_k - \vec{\bar{u}} \rangle}{\|\vec{u}_k - \vec{\bar{u}}\|} \neq 0. \tag{4.62}$$

If

$$\langle \nabla \bar{q}(\vec{\bar{u}})(t), \vec{u}_k - \vec{\bar{u}} \rangle \neq 0, \; \langle \nabla \bar{q}(\vec{u}_k)(t), \vec{u}_k - \vec{\bar{u}} \rangle \neq 0 \quad \forall k,$$

the following holds

$$\lim_{k\to\infty} \frac{M_k}{\langle \nabla \bar{q}(\vec{u}_k)(t), \vec{u}_k - \vec{\bar{u}} \rangle} = 0,$$

since

$$\begin{aligned} \lim_{k\to\infty} \frac{M_k}{\langle \bar{q}(\vec{u}_k)(t), \vec{u}_k - \vec{\bar{u}} \rangle} &= 1 - \\ \lim_{k\to\infty} \frac{\langle \nabla \bar{q}(\vec{\bar{u}})(t), \vec{u}_k - \vec{\bar{u}} \rangle}{\langle \nabla \bar{q}(\vec{u}_k)(t), \vec{u}_k - \vec{\bar{u}} \rangle} &= 1 - 1 = 0, \end{aligned}$$

where the second equality follows from (4.62). (It can be proved by dividing the numerator and denominator by $\|\vec{u}_k - \vec{\bar{u}}\| \neq 0$ and noting that $\langle \nabla \bar{q}(\vec{u}_k)(t), \vec{u}_k - \vec{\bar{u}} \rangle = \langle \nabla \bar{q}(\vec{\bar{u}})(t), \vec{u}_k - \vec{\bar{u}} \rangle +$ second order terms w.r.t. $\vec{u}_k - \vec{\bar{u}}$.)

Eventually,

$$\lim_{k\to\infty} \frac{o(t, \vec{u}_k - \vec{\bar{u}})}{|\langle \nabla \bar{q}(\vec{u}_k)(t), \vec{u}_k - \vec{\bar{u}}\rangle|} = 0 \tag{4.63}$$

since we also have (from (4.62))

$$\lim_{k\to\infty} \frac{\left[\vec{u}_k - \vec{\bar{u}}\right]^T F_k(t) \left[\vec{u}_k - \vec{\bar{u}}\right]}{|\langle \nabla \bar{q}(\vec{u}_k)(t), \vec{u}_k - \vec{\bar{u}}\rangle|} \le$$

$$\lim_{k\to\infty} \frac{\|F_k(t)\| \left\|\vec{u}_k - \vec{\bar{u}}\right\|^2}{|\langle \nabla \bar{q}(\vec{u}_k)(t), \vec{u}_k - \vec{\bar{u}}\rangle|} = 0.$$

Therefore, if $L_k^q(t) = |\langle \nabla \bar{q}(\vec{u}_k)(t), \vec{d}_k\rangle|$ then for a given $1 < \rho < \infty$, from *Lemma 4.4*, there exists a neighbourhood of $\vec{\bar{u}}$: $\mathcal{B}(\vec{\bar{u}}, \varepsilon_1)$ $(\varepsilon_1 > 0)$ such that for $\vec{u}_k \in \mathcal{B}(\vec{\bar{u}}, \varepsilon_1)$

$$\bar{q}(\vec{u}_k)(t) \le \rho \left|\left\langle \nabla \bar{q}(\vec{u}_k)(t), \vec{d}_k \right\rangle\right| = \rho L_k^q(t). \tag{4.64}$$

When $\langle \nabla \bar{q}(\vec{\bar{u}})(t), \vec{u}_k - \vec{\bar{u}}\rangle = 0$ (but $\langle \nabla \bar{q}(\vec{u}_k)(t), \vec{u}_k - \vec{\bar{u}}\rangle \ne 0$) then, from (4.59), we have $o(t, \vec{u}_k - \vec{\bar{u}}) = 1/2[\langle \nabla \bar{q}(\vec{u}_k)\ (t), \vec{u}_k - \vec{\bar{u}}\rangle - [\vec{u}_k - \vec{\bar{u}}]^T F_k(t)[\vec{u}_k - \vec{\bar{u}}]]$ and this together with the first order term (see (4.58)) result in $-1/2[\langle \nabla \bar{q}(\vec{u}_k)\ (t), \vec{\bar{u}} - \vec{u}_k\rangle + [\vec{u}_k - \vec{\bar{u}}]^T F_k(t)[\vec{u}_k - \vec{\bar{u}}]]$. The second term in this expression is negligible with respect to $\langle \nabla \bar{q}(\vec{u}_k)(t), \vec{u}_k - \vec{\bar{u}}\rangle$ (note (4.62) and (4.60)) and, according to *Lemma 4.2*, with respect to $\langle \nabla \bar{q}(\vec{u}_k)(t), \vec{d}_k\rangle$ as well.

Notice that if (4.61) holds then $\langle \nabla \bar{q}(\vec{u}_k)(t), \vec{\bar{u}} - \vec{u}_k\rangle = 0$ cannot happen. This would lead to the contradiction with (4.62).

If

$$\frac{\left|\left\langle \nabla \bar{q}(\vec{u}_k)(t), \vec{d}_k \right\rangle\right|}{\left\|\vec{d}_k\right\|} \le \nu \tag{4.65}$$

for infinitely many k, then either

$$\lim_{k\to\infty} \frac{\left|\left\langle \nabla \bar{q}(\vec{u}_k)(t), \vec{d}_k \right\rangle\right|}{\left\|\vec{d}_k\right\|} = 0, \tag{4.66}$$

or we can show that (4.62) holds.

In the first case, from (4.43),

$$\lim_{k\to\infty} \frac{|\langle \nabla \bar{q}(\vec{u}_k)(t), \vec{u}_k - \vec{\bar{u}}\rangle|}{\left\|\vec{u}_k - \vec{\bar{u}}\right\|} = 0 \tag{4.67}$$

which means that the terms: $\langle \bar{q}(\vec{u}_k)(t), \vec{u}_k - \vec{\bar{u}}\rangle$ and $o(t, \vec{u}_k - \vec{\bar{u}})$ in (4.58) are second order terms (with respect to $\vec{u}_k - \vec{\bar{u}}$). Since (4.48) holds, there exists a neighbourhood of $\bar{\vec{u}}$: $\mathcal{B}(\bar{\vec{u}}, \varepsilon_2)$ $(\varepsilon_2 > 0)$ such that, if $\vec{u}_k \in \mathcal{B}(\bar{\vec{u}}, \varepsilon_2)$, the following is true

$$\bar{q}(\vec{u}_k)(t) \leq \nu \left\|\vec{d}_k\right\| \leq \rho\nu \left\|\vec{d}_k\right\|. \tag{4.68}$$

We conclude that the following estimate holds for the case (4.65):

$$\bar{q}(\vec{u}_k)(t) \leq \rho \left|\left\langle \nabla \bar{q}(\vec{u}_k)(t), \vec{d}_k \right\rangle\right| \leq \rho\nu \left\|\vec{d}_k\right\| = \rho L_k^q(t),$$

as required.

Of course, we can prove similar results for $\bar{h}_i^1(\vec{u}_k)$, $\bar{h}_j^2(\vec{u}_k)$: there exists a neighbourhood $\mathcal{B}(\bar{\vec{u}}, \varepsilon_3)$ $(\varepsilon_3 > 0)$ such that

$$\left|\bar{h}_i^1(\vec{u}_k)\right| \leq \rho L_k^{h^1}(i) \;\; \forall i \in E, \tag{4.69}$$

$$\bar{h}_j^2(\vec{u}_k) \leq \rho L_k^{h^2}(j) \;\; \forall j \in I, \tag{4.70}$$

if $\vec{u}_k \in \mathcal{B}(\bar{\vec{u}}, \varepsilon_3)$.

Part 2). Suppose that there exists $\bar{k} < \infty$ such that $\vec{u}_{\bar{k}}$ is in the neighbourhood $\mathcal{B}(\bar{\vec{u}}, \varepsilon)$ specified in the assumption *(i)* of the theorem. Suppose also that $\vec{u}_{\bar{k}}$ is in the intersection of $\mathcal{N}$ as stated in *Theorem 4.1* and neigbourhoods $\mathcal{B}(\bar{\vec{u}}, \varepsilon_i)$, $i = 1, 2, 3$ described in *Part 1)* of the theorem.

From *Theorem 4.1*, *Lemma 4.3*, (**H6**), if only the *reduced* iterations are performed, $\{\vec{u}_k\}_{k \geq \bar{k}}$ is superlinearly convergent to a local solution of the problem $(\mathbf{P_{NLP}}(\bar{\vec{\mathbf{u}}}))$: $\bar{\vec{u}}$. ((**H6**) is needed if $T(\bar{\vec{u}})$ includes $T^v(\bar{\vec{u}})$ while *Lemma 4.3* applies to $T^c(\bar{\vec{u}})$ part of $T(\bar{\vec{u}})$.)

If $\{\vec{u}_k\}_{k \geq \bar{k}}$ is superlinearly convergent then, from *Part 1)*, we know that (4.64), (4.68), (4.69), (4.70) hold and *FD Algorithm* does not leave the *reduced* mode.

Part 3). If the *standard* iterations k_l are performed infinitely many times then, from *Theorem 3.1*, $c_k = c = const$ for k sufficiently large and

$$\bar{F}_c(\vec{u}_{k_l+1}) - \bar{F}_c(\vec{u}_{k_l}) \leq \gamma \alpha_{k_l} \sigma_{c, A_{k_l}}^{H_{k_l}}(\vec{u}_{k_l}) \;\; \forall k_l.$$

Since $\bar{F}_c(\cdot)$ is bounded from below, due to assumptions (**H1**)–(**H4**), $\{\bar{F}_c(\vec{u}_{k_l})\}$, which is a monotonically decreasing sequence, satisfies

$$\lim_{k_l \to \infty} [\bar{F}_c(\vec{u}_{k_{l+1}}) - \bar{F}_c(\vec{u}_{k_l})] = 0. \tag{4.71}$$

Since $\{\vec{u}_k\}$ is bounded, there exists a subsequence of $\{\vec{u}_{k_l}\}$ (we do not relabel) which is convergent to $\bar{\vec{u}}$. However, from (4.71), since $\bar{F}_c(\vec{u}_{k_{l+1}}) < \bar{F}_c(\vec{u}_{k_l+1})$

$$\lim_{k_l \to \infty} \alpha_{k_l} \sigma_{c, A_{k_l}}^{H_{k_l}}(\vec{u}_{k_l}) = 0.$$

This condition, as in the proof of *Theorem 4.1.1*, implies that

$$\lim_{k_l \to \infty} \sigma^{H_{k_l}}_{c_{k_l}, R_{\epsilon}, a_{k_l}}(\vec{u}_{k_l}) = 0$$

and thus $\vec{\bar{u}}$ is a stationary point of the problem $(\mathbf{P^N})$. After a finite number of iterations the subsequence $\{\vec{u}_{k_l}\}$ is in the intersection of the neighbourhood $\mathcal{N}$ specified in *Theorem 4.1* and the neighbourhood $\mathcal{B}(\vec{\bar{u}}, \varepsilon)$ specified in the assumption *(i)* of the theorem. This means that conditions specified in Step 2 of *FD Algorithm* are satisfied and the algorithm switches to the *reduced* mode. If k_l is sufficiently large $\vec{u}_k \in \mathcal{B}(\vec{\bar{u}}, \varepsilon_1) \bigcap \mathcal{B}(\vec{\bar{u}}, \varepsilon_2) \bigcap \mathcal{B}(\vec{\bar{u}}, \varepsilon_3)$, conditions for superlinear convergence are met and *FD Algorithm* does not leave the *reduced* mode as shown in *Part 2)*. This is a contradiction to the assumption that $\{k_l\}$ is infinite. ∎

Remark 4.2 *The first conclusion of the theorem holds even if we do not apply the watchdog technique specified by functions* $L_k^q(\cdot)$, $L_k^{h^1}(\cdot)$, $L_k^{h^2}(\cdot)$ *but simply switch to any locally convergent algorithm for the problem* $(\mathbf{P_{NLP}}(\vec{\bar{\mathbf{u}}}))$ *which satisfies conditions* 1)–3) *of* Remark 4.1. *However, in this case we can 'loose' many iterations of the algorithm if the local neighbourhood of* $\vec{\bar{u}}$ *defined in Theorem 4.1 has not been identified.*

Remark 4.3 *The second conclusion of the theorem holds if instead of using* Algorithm 4 *adapted for the problem* $(\mathbf{P_{NLP}}(\vec{\bar{\mathbf{u}}}))$ *we apply any superlinearly convergent algorithm which satisfies requirements of* Remark 4.1. *The method described in [74] fulfills all of these requirements.*

In order to calculate $\langle \nabla \bar{q}(\vec{u}_k)(t), \vec{d}_k \rangle$ the linearized equations corresponding to $\vec{u}_k$, $\vec{d}_k$ can be solved to obtain $y^{u_k^N, d_k^N}$ which is then used in the formula:

$$\left\langle \nabla \bar{q}(\vec{u}_k)(t), \vec{d}_k \right\rangle = q_x(t, x^{u_k^N}(t)) y^{u_k^N, d_k^N}(t).$$

This means that we do not have to evaluate gradients $\nabla \bar{q}(\vec{u}_k)(t)$ in order to perform the test in Step 5 of *FD Algorithm*. The algorithm can be implemented efficiently because there is no need to evaluate an infinite number of gradients for state constraints.

FD Algorithm requires the identification of all local maximum points and active arcs at every iteration. This can cause considerable practical problems. They can be overcome, to some extent, if we apply 'safe approach' in which doubtful points are assumed to be local maximas. Note that this strategy is closely related to the $(R_{\varepsilon,u}, \xi)$*–uniform approximation*. Note also that due to the assumption *(i)* of *Theorem 4.2* it is recommended to choose points to $T^c(\vec{u}_k)$ from the interior of an *active arc*.

Furthermore, the implementation of *FD Algorithm* significantly simplifies if state constraints are discretized *a priori*. In this case we do not have to consider separately $T^v(\vec{u}) \subset T(\vec{u})$. The number of constraints is finite and the set $T(\vec{\bar{u}})$ is identified after a finite number of iterations.

4.4 *Two–Step Superlinear Convergence*

FD Algorithm performs the *standard* iterations to guarantee global convergence and for that H_k should satisfy the condition $(\mathbf{BH^N})$. On the other hand, the reduced iterations are needed to guarantee that $\bar{\vec{u}}$ is a point of attraction for $\{\vec{u}_k\}$ and that the sequence converges fast enough if any meaningful watchdog technique, as that based on $L_k^{\cdot}(\cdot)$ is to be applied.

As we have already mentioned in *Remark 4.1* $\bar{\vec{u}}$ is a point of attraction if condition (4.41) is satisfied.

We have mentioned in *Remarks 4.1–4.3* that the method proposed in [74] fulfills all conditions for the local algorithm. However, this method is rather expensive to apply since it is based on the approximation of the Hessian matrix by finite differences. Here, we give the brief review of other alternatives for the local method. Our main aim is to propose a method which uses only a finite number of constraints of the problem $(\mathbf{P^N})$ at the expense, possibly, of the method rate of convergence.

If we assume that the reduced iterations guarantee local convergence of the algorithm, a general condition for the superlinear convergence (see, for example, [19]) is as follows:

$$\lim_{k\to\infty} \frac{\left\|\bar{Z}^T\left[H_k - \bar{H}\right]p_k\right\|}{\|p_k\|} = 0, \tag{4.72}$$

where $p_k = \vec{u}_{k+1} - \vec{u}_k$ and $\bar{Z}$ is defined as in *Theorem 4.1*.

Notice that (4.72) is satisfied when stronger condition holds:

$$\lim_{k\to\infty} \left\|\bar{Z}^T\left[H_k - \bar{H}\right]\right\| = 0$$

(c.f. *Theorem 4.1*).

If the rank–two Powell–symmetric–Broyden (PSB) formula is applied:

$$\begin{aligned} H_{k+1} &= H_k + \frac{1}{\|p_k\|^2}\left[(q_k - H_k p_k)p_k^T + p_k(q_k - H_k p_k)^T\right] - \\ &\quad \frac{(q_k - H_k p_k)^T p_k}{\|p_k\|^2} p_k p_k^T \end{aligned}$$

where q_k is defined by

$$q_k = \nabla_{\vec{u}}\bar{L}(\vec{u}_{k+1}, \lambda_k) - \nabla_{\vec{u}}\bar{L}(\vec{u}_k, \lambda_k), \tag{4.73}$$

and $\bar{L}(\vec{u}_k, \lambda_k) = \bar{F}_0(\vec{u}_k) + \sum_{i\in E^k}(\lambda_{i,1}^{1,k} - \lambda_{i,2}^{1,k})\bar{h}_i^1(\vec{u}_k) + \sum_{j\in I^k}\lambda_j^{2,k}\bar{h}_j^2(\vec{u}_k) + \sum_{t\in T^k}\lambda^k(t)\bar{q}(\vec{u}_k)(t)$, then $\{H_k\}$ will satisfy (4.72) ([19]). Here, $\{\lambda_{i,1}^{1,k}\}$, $\{\lambda_{i,2}^{1,k}\}$, $\{\lambda_j^{2,k}\}$, $\{\lambda^k(t)\}$ are the Lagrange multipliers of $\mathbf{P}^{\mathbf{H_k}}_{\mathbf{c_k},\mathbf{A_k}}(\vec{\mathbf{u}}_\mathbf{k})$ and $E^k = E^+(\vec{u}_k)\bigcup E^-(\vec{u}_k)$, $I^k = I(\vec{u}_k)$, $T^k = T(\vec{u}_k)$ if *FD Algorithm* is in the standard mode, or $E^k = E_r^+\bigcup E_r^-$, $I^k = I_r$, $T^k = T_r^c\bigcup\{m_l^t(\vec{v}, \vec{u}_k) : l \in \mathcal{K}_r^s\}$ otherwise.

Although H_k generated by the PSB formula are uniformly bounded, as are their inverses H_k^{-1}, they do not fulfill (4.41) ([19], [56]). As a result equations (4.35)–(4.36) should be used instead to calculate $\vec{d}_k$.

The condition (4.72) can be rewritten as

$$\lim_{k\to\infty}\left\{\frac{\bar{Z}^T\left[H_k-\bar{H}\right]\bar{Z}p_k^Z}{\|p_k\|}+\frac{\bar{Z}^T\left[H_k-\bar{H}\right]\bar{Y}p_k^Y}{\|p_k\|}\right\}=0.$$

Here, $\bar{Y}$ is defined by rows of the matrix $\bar{A}$ and p_k^Z, p_k^Y are coordinates of p_k in the null and the range spaces respectively:

$$p_k=\bar{Z}p_k^Z+\bar{Y}p_k^Y.$$

Slower convergence, but still sufficient to construct a satisfactory watchdog technique, is obtained if

$$\lim_{k\to\infty}\frac{\left\|\bar{Z}^T\left[H_k-\bar{H}\right]\bar{Z}p_k^Z\right\|}{\|p_k\|}=0. \tag{4.74}$$

If the sequence $\{p_k\}$ approaches zero in such a way that

$$\lim_{k\to\infty}\frac{\bar{Y}p_k^Y}{\|p_k\|}=0$$

then (4.74) implies (4.72) and the superlinear convergence results.

The slower convergence is the two–step superlinear convergence as described in the following definition.

Definition 4.3 *The sequence $\{\vec{u}_k\}$ is two–step superlinearly convergent to the point $\bar{\vec{u}}$ if there exists a sequence $\{\alpha_k\}$ converging to zero such that*

$$\left\|\vec{u}_{k+2}-\bar{\vec{u}}\right\|\leq\alpha_k\left\|\vec{u}_k-\bar{\vec{u}}\right\|\ \ k=0,1,\ldots$$

The property analogous to (4.33) holds for a sequence two–step superlinearly convergent:

$$\lim_{k\to\infty}\frac{\left\|\vec{d}_{k+1}+\vec{d}_k\right\|}{\left\|\vec{u}_k-\bar{\vec{u}}\right\|}=1. \tag{4.75}$$

The two–step superlinear convergence is fast enough from practical point of view. Moreover, one can expect that it is much easier to construct $\{H_k\}$ in such a way that H_k satisfy ($\mathbf{BH}^{\mathrm{N}}$) and at the same time condition (4.74) is fulfilled. Therefore, the question is whether new functions similar to $L_k^{\cdot}(\cdot)$ could be proposed together with the watchdog technique based on them.

The first observation is that *Lemma 4.2* is also valid for two–step convergent sequences.

Lemma 4.4 *If the sequence* $\vec{u}_{k+2} = \vec{u}_k + \vec{d}_k + \vec{d}_{k+1}$ *is superlinearly convergent to* $\bar{\vec{u}}$, $\{\vec{z}_k\}$ *is a convergent sequence and the following holds:*

$$\vec{d}_k + \vec{d}_{k+1} \neq 0, \qquad \vec{u}_k - \bar{\vec{u}} \neq 0 \ \forall k,$$

then

$$\lim_{k\to\infty} \phi_k = \lim_{k\to\infty} \gamma_k,$$

where ϕ_k *is the angle between* $\vec{u}_k - \bar{\vec{u}}$ *and* $\vec{z}_k$ *and* γ_k *the angle between* $\vec{d}_k + \vec{d}_{k+1}$ *and* $\vec{z}_k$.

Proof.

The proof of the lemma is the simple modification of the proof of *Lemma 4.2.* ■

Lemma 4.4 suggests the following modifications of functions $L_k^{\cdot}(\cdot)$:

$$\bar{L}_k^q(t) =$$
$$\begin{cases} |\langle \nabla\bar{q}(\vec{u}_{k-1})(t), \vec{d}_{k-1} + \vec{d}_k\rangle| & \text{if } \dfrac{|\langle \nabla\bar{q}(\vec{u}_{k-1})(t), \vec{d}_{k-1} + \vec{d}_k\rangle|}{\|\vec{d}_{k-1} + \vec{d}_k\|} > \nu \\ \nu\|\vec{d}_{k-1} + \vec{d}_k\| & \text{if } \dfrac{|\langle \nabla\bar{q}(\vec{u}_{k-1})(t), \vec{d}_{k-1} + \vec{d}_k\rangle|}{\|\vec{d}_{k-1} + \vec{d}_k\|} \leq \nu \end{cases}$$
$$\bar{L}_k^{h^1}(i) =$$
$$\begin{cases} |\langle \nabla\bar{h}_i^1(\vec{u}_{k-1}), \vec{d}_{k-1} + \vec{d}_k\rangle| & \text{if } \dfrac{|\langle \nabla\bar{h}_i^1(\vec{u}_{k-1}), \vec{d}_{k-1} + \vec{d}_k\rangle|}{\|\vec{d}_{k-1} + \vec{d}_k\|} > \nu \\ \nu\|\vec{d}_{k-1} + \vec{d}_k\| & \text{if } \dfrac{|\langle \nabla\bar{h}_i^1(\vec{u}_{k-1}), \vec{d}_{k-1} + \vec{d}_k\rangle|}{\|\vec{d}_{k-1} + \vec{d}_k\|} \leq \nu \end{cases}$$
$$\bar{L}_k^{h^2}(j) =$$
$$\begin{cases} |\langle \nabla\bar{h}_j^2(\vec{u}_{k-1}), \vec{d}_{k-1} + \vec{d}_k\rangle| & \text{if } \dfrac{|\langle \nabla\bar{h}_j^2(\vec{u}_{k-1}), \vec{d}_{k-1} + \vec{d}_k\rangle|}{\|\vec{d}_{k-1} + \vec{d}_k\|} > \nu \\ \nu\|\vec{d}_{k-1} + \vec{d}_k\| & \text{if } \dfrac{|\langle \nabla\bar{h}_j^2(\vec{u}_{k-1}), \vec{d}_{k-1} + \vec{d}_k\rangle|}{\|\vec{d}_{k-1} + \vec{d}_k\|} \leq \nu, \end{cases} \tag{4.76}$$

where $\nu > 0$.

Then, Step 5 of *FD Algorithm* is replaced by

5′. If **fop** is *reduced* and

$$\begin{aligned} \bar{q}(\vec{u}_{k-1})(t) &> \varrho\bar{L}_k^q(t) \ \text{ for some } t \in R_{\varepsilon,\vec{u}_k}, \text{ or} \\ \bar{h}_i^1(\vec{u}_{k-1}) &> \varrho\bar{L}_k^{h^1}(i) \ \text{ for some } i \in E, \text{ or} \\ \bar{h}_j^2(\vec{u}_{k-1}) &> \varrho\bar{L}_k^{h^2}(j) \ \text{ for some } j \in I, \end{aligned}$$

set **fop** to *standard*, substitute k_s for k, $\vec{v}$ for $\vec{u}_k$ and go to Step 3.

We can prove the following theorem.

Theorem 4.3 *If the assumptions of* Theorem 4.2 *are satisfied with the exception that condition (4.34) is replaced by (4.74) and the assumption (ii) by*

(ii') $\bar{L}^q_k(\cdot)$, $\bar{L}^{h^1}_k(\cdot)$, $\bar{L}^{h^2}_k(\cdot)$ are defined by (4.76)

then

1) *after a finite number of iterations a fixed number of constraints is used to define the sequence $\{\vec{u}_k\}$,*

2) *$\{\vec{u}_k\}$ is two-step superlinearly convergent to a stationary point of the problem ($\mathbf{P^N}$).*

Proof.

The proof is the same as the proof of *Theorem 4.2.* ■

The remaining question is whether the sequence $\{H_k\}$ which satisfies ($\mathbf{BH^N}$) and (4.74) can be constructed. It is suggested in [19] (see also [44], [18]) that this would be really the case. It would be possible to provide positive definite updates which would also satisfy (4.74).

We have mentioned in *Remark 4.1* and *Remark 4.3* that, in the reduced mode, *Algorithm 4* can be replaced by any superlinearly convergent algorithm which fulfills requirements of *Remark 4.1.* We can therefore apply algorithms which approximate not the Hessian of the Lagrangian but

$$\bar{Z}^T \bar{H} \bar{Z}. \tag{4.77}$$

A two-step superlinearly convergent algorithm based on the approximation of (4.77) is described in [30], while superlinerly convergent method is presented in [17].

Algorithms we have suggested for the reduced mode iterations require special techniques for solving (4.35)–(4.36) different from those we can apply to $\mathbf{P}^{\mathbf{H_k}}_{\mathbf{c_k},\mathbf{A_k}}(\vec{\mathbf{u}}_\mathbf{k})$ (because H_k, in general, do not satisfy ($\mathbf{BH^N}$)). Therefore it is natural to ask the question whether the watchdog technique discussed in this chapter could require even slower rate of convergence than the two-step superlinear. Let us notice that the watchdog technique is possible due to relations (4.33), (4.75) which hold for superlinearly convergent subsequences. If only linear convergence can be guaranteed for a sequence $\{\vec{u}_k\}$:

$$\left\|\vec{u}_{k+1} - \bar{\vec{u}}\right\| \leq \alpha_k \left\|\vec{u}_k - \bar{\vec{u}}\right\| \tag{4.78}$$

where $\alpha_k \to \alpha \in (0,1)$, then we also have

$$1-\alpha \leq \lim_{k\to\infty} \frac{\|\vec{u}_{k+1}-\vec{u}_k\|}{\|\vec{u}_k-\bar{\vec{u}}\|} \leq 1+\alpha. \tag{4.79}$$

We conjecture that (4.79) should lead to a viable watchdog technique as stated in *Theorem 4.2* under the condition that ϱ (in *FD Algorithm*) is sufficiently greater than one. Unfortunately we haven't proved that and we anticipate that the proofs of *Lemma 4.2* and *Theorem 4.2* would have to be modified significantly.

Below we present some numerical results for the case when $\{H_k\}$ are updated by the BFGS scheme with the Powell's modifications ([88]).

$$\begin{aligned} H_{k+1} &= H_k - \frac{H_k p_k p_k^T H_k}{p_k^T H_k p_k} + \frac{r_k^T r_k}{p_k^T r_k}, & (4.80)\\ r_k &= \theta_k q_k + (1-\theta_k) H_k p_k, & (4.81)\\ p_k &= \vec{u}_{k+1} - \vec{u}_k, & (4.82)\\ q_k &= \nabla_{\vec{u}} \bar{L}(\vec{u}_{k+1}, \lambda_k) - \nabla_{\vec{u}} \bar{L}(\vec{u}_k, \lambda_k), & (4.83) \end{aligned}$$

where $\bar{L}(\vec{u}_k, \lambda_k)$ is defined as in (4.73) and

$$\theta_k = \begin{cases} 1 & \text{if } p_k^T q_k \geq 0.2 p_k^T H_k p_k \\ \dfrac{0.8 p_k^T H_k p_k}{p_k^T H_k p_k - p_k^T q_k} & \text{if } p_k^T q_k < 0.2 p_k^T H_k p_k. \end{cases} \tag{4.84}$$

H_0 can be any symmetric positive definite matrix, although the choice of $H_0 = \text{diag}(\text{sqrt}(1/N))$ is recommended.

We can show that H_k updated in this way are always positive definite ([12]). On the other hand, the only convergence result which applies to this update states that if $\{\vec{u}_k\}$ *converges* and $\{\|H_k\|\}$, $\{\|Z_k^T H_k Z_k\|\}$ are bounded (Z_k is defined analogously to $\bar{Z}$) then the rate of convergence is R–superlinear (see [82] for the definition of R–superlinear convergence). R–superlinear convergence does not even imply linear convergence as defined in (4.78). However, in practice, the Powell's update works well and usually at least a linear rate of convergence is achieved.

4.5 Numerical Experiments

FD Algorithm was tested on several optimal control problems with state constraints. Among them were two problems reported in Chapter 4: the brachistochrone problem (Example 1) and the crane problem (Example 2).

The solution to the brachistochrone problem has an active arc in two thirds of the horizon. The solution to the crane problem has active arcs reduced to single points (only local maximum points). The results for the control discretization equal $N = 20$ and $N = 100$ are presented in Table 5.1, where ITN is the number of iterations, ITR is the number of iterations with **fop** set to *reduced*, LMS is $|T(\vec{u})| + |E(\vec{u})| + |I(\vec{u})|$. We applied $(R^d_{\varepsilon,u_k}, \xi^j_k)$–*uniform approximations* as described on p. 69 with $\xi^0_k = 1/(2N)$. The relatively dense approximation of state constraints was chosen to verify behaviour of the watchdog technique introduced in this chapter.

PROBLEM	ITN	ITR	LMS	$\max_{t \in T} q(t, \bar{x}(t))$
Brachistochrone				
N=20	8	2	14	$1.0 \cdot 10^{-8}$
N=100	17	9	64	$3.4 \cdot 10^{-6}$
Crane				
N=20	9	3	7	$4.9 \cdot 10^{-8}$
N=100	8	1	7	$2.6 \cdot 10^{-7}$

TABLE 5.1. Performance of *FD Algorithm.*

The efficiency of *FD Algorithm* strongly depends on an initial approximation set A^1_k. The following choice is recommended. Set A^1_k to $(R_{\varepsilon,\bar{u}_k}, \xi)$–*uniform approximation* with $\xi = 1/N$. It usually does not require the update of ξ.

To explain this consider the brachistochrone example with a single active arc at the solution $(\bar{x}, \bar{u})$:

$$\bar{x}_2(t) - \bar{x}_1(t)\tan(\theta) - h = 0, \quad t \in [t_{en}, t_{ex}], \; t_{en} > t_{ex}. \qquad (4.85)$$

Differentiation of (4.85) with respect to time leads to the relation

$$\sqrt{2g\bar{x}_2(t)}\sin(\bar{u}(t)) - \sqrt{2g\bar{x}_2(t)}\cos(\bar{u}(t))\tan(\theta) = 0, \quad t \in [t_{en}, t_{ex}],$$

which, under the assumption $\bar{x}_2(t) \neq 0$, $t \in [t_{en}, t_{ex}]$, implies that $\bar{u}(t) = \theta$, $t \in [t_{en}, t_{ex}]$. Therefore, we can expect that only one point from each $[t_j, t_{j+1}] \bigcap R_{\varepsilon,\bar{u}_k}$, where $t_{j+1} - t_j = 1/N$, would be needed in direction finding subproblems.

FD Algorithm was also applied to optimal control problems defined by large–scale differential–algebraic equations. The results of these tests are discussed in Chapter 6. The fact that $(R_{\varepsilon,\bar{u}_k}, \xi^0_k)$–*uniform approximation* of state constraints with $\xi^0_k = 1/N$ did not require further updates for all problems considered in the next chapter is of great importance. The evaluation of adjoint equations needed for the reduced gradients is expensive for problems described by large–scale differential–algebraic equations.

Note that this choice of ξ_k^0, for the examples analysed in Chapter 6, resulted in maximum sets of linearly independent constraints. This follows from the special structure of vectors $\nabla\bar{q}(\vec{u})(\cdot)$ as shown in (4.6)–(4.7).

We should also emphasize the fact that substituting controls (measurable functions in general) by piecewise constant approximations qualitatively changes an optimal control problem especially when state constraints are present in its formulation.

Consider the optimal control problem

$$\min_u \phi(x_1(1)) \tag{4.86}$$

subject to the constraints

$$\begin{aligned} \dot{x}_1(t) &= x_2(t), && (4.87)\\ \dot{x}_2(t) &= u(t), && (4.88)\\ x_2(t) - f(t) &\leq 0, \quad t \in [0,1]. && (4.89) \end{aligned}$$

Assume that $(\bar{x}, \bar{u})$ is the solution to (4.86)–(4.89) and that there is an active arc for (4.89) at the solution:

$$\bar{x}_2(t) - f(t) = 0, \quad t \in [t_{en}, t_{ex}], \ t_{ex} > t_{en}. \tag{4.90}$$

Differentiation of (4.90) with respect to time results in the equation

$$\bar{u}(t) - \dot{f}(t) = 0, \quad t \in [t_{en}, t_{ex}].$$

If f is not constant on $[t_{en}, t_{ex}]$ then we cannot expect that (4.86)–(4.89) with piecewise constant controls will have an active arc at a solution although touch points can be present. This in fact simplifies our control problem (at the expense of the optimal value of the objective function) because active arcs would result in a nonlinear programming problem with an infinite number of active constraints at its solution. However, the discretization of states (and thus state constraints) will not eliminate redundant constraints if $\dot{f} \equiv 0$ on $[t_{en}, t_{ex}]$.

If decision variables are parameters w (c. f. 4.3.1) very little, in general, can be said about the structure of vectors $\nabla\bar{q}(w)(\cdot)$. This means that problems with parameters w, instead of control functions $\vec{u}$, as decision variables are more difficult to solve. However, the second order algorithm presented in this chapter can still be applied to these problems.

5 Concluding Remarks

In this chapter we present a second order method for optimal control problems with state constraints. The notable feature of the method is the fact that, under some regularity assumptions, only a finite number of constraints is needed in direction finding subproblems. Furthermore, gradients of these constraints are linearly independent and this guarantees that the sequence generated by the method is superlinearly convergent to a local solution.

The method uses similar scheme for the approximation of state constraints as the first order method discussed in Chapters 3–4. This scheme increases the number of approximation points in accordance with the directional minimization. If the directional minimization does not require small steps it means that the approximation to state constraints is adequate. On the other hand, small steps suggest that directions generated in the directional minimization phase of the method are not directions of descent due to poor representation of state constraints. The second order method described in this chapter has the property of choosing the finite and adequate representation of state constraints and thus, after a finite number of iterations, there is no need for increasing the number of the approximation points.

6

Runge–Kutta Based Procedure for Optimal Control of Differential — Algebraic Equations

We introduce the discretization of state trajectories and show that we can solve large-scale optimal control problems.

1 Introduction

In this chapter we consider the optimal control problem ($\mathbf{P_{DAE}}$) described by the fully implicit differential–algebraic equations:

$$\min_u \phi(x(1))$$

subject to the constraints

$$\begin{aligned} F(t, \dot{x}(t), x(t), y(t), u(t)) &= 0 \text{ a.e. on } T,\ x(0) = x_0 && (1.1)\\ q(t, x(t)) &\leq 0\ \forall t \in T && (1.2)\\ h_i^1(x(1)) &= 0\ \forall i \in E && (1.3)\\ h_j^2(x(1)) &\leq 0\ \forall j \in I && (1.4)\\ u \in \mathcal{U} &= \{u:\ u(t) \in \Omega \text{ a.e. on } T\}. && (1.5) \end{aligned}$$

Here, $x(t) \in \mathcal{R}^{nd}$, $y(t) \in \mathcal{R}^{na}$, $u(t) \in \mathcal{R}^{m}$, $n = nd + na$ and Ω is a convex compact set. We assume that for any x_0, $u \in \mathcal{U}$ there exists a unique solution to (1.1), (x^u, y^u). We call x a *differential* state and y an *algebraic* state.

This chapter is written differently from the previous chapters. We do not concentrate on precise mathematical description of ($\mathbf{P_{DAE}}$). Instead, we present the implementation of an implicit Runge–Kutta integration procedure and its application to algorithms for optimal control problems. The main reason for not very detailed description lies in our inadequate understanding of optimal control problems defined by differential–algebraic equations. We still lack, for example, general conditions under which a solution to the problem ($\mathbf{P_{DAE}}$) exists ([22]), also optimality conditions, necessary and sufficient, are not stated for a general problem. To appreciate complexity of the problem ($\mathbf{P_{DAE}}$) see, for instance, [97]. Fortunately, for index one systems (see the description below), which describe the majority of practical problems, we can apply procedures introduced in Chapters 3–5.

Assume that partial derivatives of F with respect to $\dot{x}$, x , y and u exist and are continuous and that

$$\det\left[F_{\dot{x}}(t,\dot{x}(t),x(t),y(t),u(t)),F_y(t,\dot{x}(t),x(t),y(t),u(t))\right]\neq 0 \tag{1.6}$$

for any $\dot{x}(t)$, $x(t)$, $y(t)$, $u(t)$ and t. We say that if (1.6) is satisfied, system (1.1) has index one ([22]).

If a system has index one, from the Implicit Function Theorem, there exist functions Φ_1 and Φ_2 such that

$$\dot{x}(t) = \Phi_1(t,x(t),u(t)) \tag{1.7}$$

$$y(t) = \Phi_2(t,x(t),u(t)) \text{ a.e. on } T. \tag{1.8}$$

Furthermore,

$$F(t,\Phi_1(t,x(t),u(t)),x(t),\Phi_2(t,x(t),u(t)),u(t))=0 \text{ a. e. on } T. \tag{1.9}$$

In particular,

$$F(\Phi_1(0,x_0,u(0)),x_0,\Phi_2(0,x_0,u(0)),u(0),0)=0,$$

thus initial conditions for (1.1), given by x_0, are consistent.

Due to our assumption that initial conditions for (1.1) are consistent and that for any $u\in\mathcal{U}$ there exists a unique solution to (1.1), (x^u,y^u), we can state the problem ($\mathbf{P_{DAE}}$) as the problem (**P**). Therefore, any method discussed in Chapters 3–5 can be used to solve the problem ($\mathbf{P_{DAE}}$). However, in order to solve problem (1.3)–(1.6), stated as the problem (**P**), a procedure for calculating $\nabla\tilde{F}_0(u)$, $\nabla\tilde{h}_i^1(u)$, $i\in E$, $\nabla\tilde{h}_j^2(u)$, $j\in I$, $\tilde{q}(u)(t)$, $t\in T$ has to be provided.

If the dimension of a state vector (x,y) is small, one way of solving the problem ($\mathbf{P_{DAE}}$) is to substitute both state and control functions by their polynomial approximations described by a finite number of parameters. Then, instead of solving the problem ($\mathbf{P_{DAE}}$) we cope with a finite dimensional nonlinear programming problem which contains equality constraints corresponding to system equations (1.1). The nonlinear programming problem can be solved by any constrained optimization procedure although the reduced gradient version of it can be particularly efficient due to the fact that there are relatively few degrees of freedom ([1],[17],[80]). This approach, in general, does not work very well for problems with many differential–algebraic equations because the number of parameters to be optimized is very large although reasonably good results can be obtained especially for problems without steep profiles ([78]).

The next approach to the problem ($\mathbf{P_{DAE}}$) does not assume that a state vector is parametrized. The problem (1.3)–(1.6) is solved by introducing sensitivity equations for evaluation of the gradients $\nabla\tilde{F}^0(u)$, $\nabla\tilde{h}_i^1(u)$,

$i \in E$, $\nabla \tilde{h}_j^2(u)$, $j \in I$, $\nabla \tilde{q}(u)(t)$, $t \in T$. If controls are approximated by piecewise constant functions, then parameters of these approximations $p = (p_1, \ldots, p_l)$ will define $x^p(t)$, $y^p(t)$, $t \in T$ which will have the following sensitivity equations corresponding to equations (1.1):

$$F_{\dot{x}}(t)\dot{s}_j^d(t) + F_x s_j^d(t) + F_y(t)s_j^a(t) + F_{p_j}(t) = 0, \text{ a.e. on } T, \quad j = 1, \ldots, l \tag{1.10}$$

where

$$s_j^d(t) = \frac{\partial x^p(t)}{\partial p_j}^T, \quad s_j^a(t) = \frac{\partial y^p(t)}{\partial p_j}^T$$

are sensitivity vectors and $F_{\dot{x}}(t)$, for example, means $F_{\dot{x}}(t, \dot{x}^p(t), x^p(t), y^p(t), p)$. Then, $\hat{h}_i^1(p) = h_i^1(x^p(1))$ (for example) will have partial derivatives: $(\hat{h}_i^1)_{p_j}(p) = (h_i^1)_x(x^p(1))s_j^d(1)$, $j = 1, \ldots, l$.

This approach has the following features.

(i) It is conservative in the sense that it provides $s_j^d(t)$ for every $t \in T$ while we need these values only at times where the gradients of constraints are evaluated. For example, if only terminal constraints (1.3)–(1.4) are present we will need $s_j^d(1)$, $j = 1, \ldots, l$. This implies that

(ii) the number of sensitivity equations substantially exceeds the number of original equations—for each parameter p_j, n linear time varying equations have to be solved.

The approach based on sensitivity equations can cope well with moderate size problems—however the number of sensitivity equations precludes its efficient application to systems described by large–scale differential–algebraic equations. The approach can still be used for large systems, if the number of parameters l is limited, but then calculated control profiles are only very crude approximations to solutions of the problem $(\mathbf{P_{DAE}})$.

Both approaches described above are based on finite dimensional approximations of control functions (the first also on finite dimensional approximations of state functions) thus it is tempting to extend optimization methods designed for control problems described by ordinary differential equations (such as those described in Chapters 3–5) to the problem $(\mathbf{P_{DAE}})$.

The first formula we have to establish is that concerning reduced gradients. From (1.7)–(1.9) we have:

$$F_{\dot{x}}(t)(\Phi_1)_x(t) + F_x(t) + F_y(t)(\Phi_2)_x(t) = 0 \tag{1.11}$$

$$F_{\dot{x}}(t)(\Phi_1)_u(t) + F_u(t) + F_y(t)(\Phi_2)_u(t) = 0 \tag{1.12}$$

and all functions $F_{\dot{x}}(t)$, $F_x(t)$, $F_y(t)$, $F_u(t)$ are evaluated at $(t, \dot{x}^u(t), x^u(t), y^u(t), u(t))$ and $(\Phi_1)_x(t)$, $(\Phi_2)_x(t)$, $(\Phi_1)_u(t)$, $(\Phi_2)_u(t)$ at $(t, x^u(t), u(t))$.

Knowing that condition (1.6) is satisfied we can write

$$\begin{bmatrix} (\Phi_1)_x(t) \\ \\ (\Phi_2)_x(t) \end{bmatrix} = -\left[F_{\dot{x}}(t), F_y(t)\right]^{-1} F_x(t)$$

and

$$\begin{bmatrix} (\Phi_1)_u(t) \\ \\ (\Phi_2)_u(t) \end{bmatrix} = -\left[F_{\dot{x}}(t), F_y(t)\right]^{-1} F_u(t).$$

As we have shown in Chapter 2 the reduced gradient for the functional $\phi(x^u(1))$ and equations (1.7) can be calculated with the help of adjoint equations:

$$\begin{aligned} p_c(1) &= \phi_x(x^u(1))^T \\ \dot{p}_c(t) &= -(\Phi_1)_x(t)^T p_c(t), \text{ a. e. on } T. \end{aligned}$$

If we define r_c in the following way

$$\left[F_{\dot{x}}(t), F_y(t)\right]^T r_c(t) = \begin{bmatrix} p_c(t) \\ 0 \end{bmatrix} \tag{1.13}$$

these equations can be written as

$$\begin{aligned} p_c(1) &= \phi_x(x^u(1))^T \\ \dot{p}_c(t) &= F_x(t)^T r_c(t), \text{ a. e. on } T. \end{aligned} \tag{1.14}$$

(In order to obtain (1.14) transpose equations (1.11), move $F_x(t)^T$ on the right–hand side, multiply both sides, on the left, by $r_c(t)$ and take into account (1.13).)

The reduced gradient is then calculated according to the formula

$$\nabla \tilde{F}_0(u)(t) = (\Phi_1)_u(t)^T p_c(t),$$

thus

$$\nabla \tilde{F}_0(u)(t) = -F_u(t)^T r_c(t).$$

The implementation of these formulas is hampered by the following drawbacks:

(i) Jacobians for system and adjoint equation integration procedures can be evaluated at different times therefore, in general, we cannot use sparse LU factors of Jacobians ([39]) from system equation integration while solving adjoint equations,

(ii) if system equations are integrated with low accuracy, which is the typical situation for large–scale equations (1.1), then the reduced gradients will be inaccurately calculated due to inconsistency of discrete time representation of system and adjoint equations.

If we use the reduced gradients described above then an integration procedure based on the backward differential formula (BDF) can be used to integrate the system and adjoint equations. The fact that control functions are typically substituted by their piecewise constant approximations as described in Chapter 4 (thus discontinuous functions) can cause deterioration in the otherwise satisfactory performance of BDF codes for large–scale systems of equations ([22]).

2 The Method

The difficulties of applying the three approaches described above are overcome in the approach proposed here. It is based on a discrete time approximation of system (1.1) (thus it is close to the first approach) and is designed for a discrete time representation of the problem (**P**) (thus it is close to the second and third approaches as well—all functionals defining ($\mathbf{P_{DAE}}$) are treated as functionals of u only). The method we introduce assumes that system equations (1.1) have been discretized by an implicit Runge–Kutta procedure.

Before presenting the discrete time equations we have to distinguish two different types of discretization:

(C1) discretization of controls: the ordered set $\mathcal{D}_u = \{i_0, i_1, \ldots, i_{N-1}\}$

(C2) discretization of states: the ordered set $\mathcal{D}_s = \{k_0, k_1, \ldots, k_{N_s}\}$.

The discretization of controls is an *a priori* discretization while that of states is defined by integration steps.

We denote h^{k_l} by

$$h^{k_l} := \sum_{j=0}^{l} h(k_j), \tag{2.1}$$

where $k_j \in \mathcal{D}_s$ and $h(k_j)$ are integration steps, and we impose the following condition on these discretizations

$$l \in \mathcal{D}_u \Longrightarrow l \in \mathcal{D}_s. \tag{2.2}$$

(2.2) follows from the fact that at times h^{i_l}, where $i_l \in \mathcal{D}_u$, $\dot{x}$ can be discontinuous thus these times must coincide with the integration steps.

We denote by u^N a piecewise constant approximation to a control u defined on N equal subintervals as described in Chapter 4. To simplify the further notation we write

$$u(j) := u^N(h^j), \quad j \in \mathcal{D}_s. \tag{2.3}$$

Moreover, we write $\vec{u} = (u(i_0), \ldots, u(i_{N-1}))$, $i_k \in \mathcal{D}_u$, $k = 0, \ldots, N-1$, and also $\vec{u} = (u(0), \ldots, u(N-1))$.[1] To simplify the presentation of the integration procedure we assume that $\mathcal{D}_s$ can be regarded as $\{0, 1, \ldots, N_s\}$.

2.1 Implicit Runge–Kutta Methods

The integration scheme we apply to equations (1.1) is an implicit Runge–Kutta algorithm. Before introducing discrete time equations resulting from the application of the scheme we would like to discuss, very briefly, Runge–Kutta methods. They have been originally conceived for the numerical solution of ordinary differential equations $\dot{x}(t) = f(x(t))$. From an approximation $x(k)$ of the solution at h^k these one–step methods construct an approximation $x(k+1)$ at $h^{k+1} = h^k + h(k)$ via the following formulas[2]

$$x(k+1) = x(k) + h(k) \sum_{i=1}^{s} b_i x_i^{'}(k+1) \tag{2.4}$$

where

$$x_i^{'}(k+1) = f(x_i(k+1)), \; i = 1, \ldots, s \tag{2.5}$$

with internal stages $x_i(k+1)$ defined by

$$x_i(k+1) = x(k) + h(k) \sum_{j=1}^{s} a_{ij} x_j^{'}(k+1), \; i = 1, \ldots, s. \tag{2.6}$$

Here, b_i, a_{ij} are the coefficients which determine the method and s is the number of stages. If $a_{ij} = 0$ for $i \le j$ we compute internal stages $x_1(k)$, $\ldots, x_s(k)$ one after the other from (2.6). Such methods are called explicit. The others, for which (2.5)–(2.6) constitute a nonlinear system of algebraic equations for the internal stages, are called implicit.

The equation (2.5) is of the same form as the differential equation $\dot{x}(t) = f(x(t))$ itself while equations (2.4) and (2.6) depend on the method coefficients and the stepsize $h(k)$, but are independent of the special form of

[1]Notice that $u^{N,j} = u(j)$ according to the notation of Chapter 4.

[2]Notice that we apply a different convention than in [54], for example, to describe an implicit Runge–Kutta method—our convention emphasizes that $x_i^{'}(k+1)$, $x_i(k+1)$ and $x(k+1)$ are calculated by the same iteration process.

the differential equation. This suggests an extension of the Runge–Kutta method to differential–algebraic equations

$$f(\dot{x}(t), x(t), y(t)) = 0$$

by defining $x(k+1)$ as the solution of the system (2.4), (2.6) and

$$\begin{aligned} f(x_i'(k+1), x_i(k+1), y_i(k+1)) &= 0, \; i = 1, \ldots, s \\ f(x'(k+1), x(k+1), y(k+1)) &= 0, \end{aligned}$$

where the last equation is needed for the approximation $y(k+1)$ of y at h^{k+1}.

As it was noted in [54] explicit Runge–Kutta methods are not directly suited for this approach because for differential–algebraic equations some components of $x_i'(k+1)$ have to be determined from (2.6) and for that the matrix $A = \{a_{ij}\}$ must be invertible (the system in question is, for example, $\dot{x}_1 = x_2$, $x_1 = 0$).

To describe briefly some classes of implicit Runge–Kutta methods we introduce

$$c_i = \sum_{j=1}^{s} a_{ij}, \; i = 1, \ldots, s$$

and the conditions

$$B(p): \quad \sum_{i=1}^{s} b_i c_i^{k-1} = \frac{1}{k}, \; k = 1, \ldots, p \tag{2.7}$$

$$C(q): \quad \sum_{j=1}^{s} a_{ij} c_j^{k-1} = \frac{c_i^k}{k}, \; k = 1, \ldots, q \;\; \forall i \tag{2.8}$$

$$D(r): \quad \sum_{i=1}^{s} b_i c_i^{k-1} a_{ij} = \frac{b_j}{k}(1 - c_j^k), \; k = 1, \ldots, r \;\; \forall j. \tag{2.9}$$

Condition $B(p)$ means that the quadrature formula with weights $b_1, \ldots, b_s$ and nodes $c_1, \ldots, c_s$ integrates polynomials up to degree $p-1$ exactly on the interval $[0, 1]$. Condition $C(q)$ means that the quadrature formula with weights $a_{i1}, \ldots, a_{is}$ integrates polynomials up to degree at least $q-1$ on the interval $[0, c_i]$ for each i. Below we list properties of coefficients of some classical methods

Gauss:	$B(2s)$, $C(s)$, $D(s)$
Radau IA:	$B(2s-1)$, $C(s-1)$, $D(s)$, $c_1 = 0$
Radau IIA:	$B(2s-1)$, $C(s)$, $D(s-1)$, $c_s = 1$, $b_i = a_{si}$
Lobatto IIIA:	$B(2s-2)$, $C(s)$, $D(s-2)$, $c_1 = 0$, $c_s = 1$, $b_i = a_{si}$
Lobatto IIIC:	$B(2s-2)$, $C(s-1)$, $D(s-1)$, $c_1 = 0$, $c_s = 1$, $b_i = a_{si}$.

The condition $b_i = a_{si}$ means that $x(k+1) = x_s(k+1)$ and this is a very favorable property of the method because it implies fewer nonlinear equations to be solved at every step. For Lobatto IIIA methods (the trapezoidal rule belongs to them) the first row of the matrix $A = \{a_{ij}\}$ is identically 0, so that A is not invertible. However $\{a_{ij}\}_{i,j=2}^{s}$ is invertible and $b_i = a_{si}$ for all i and the method is well–defined.

Other classical implicit Runge–Kutta methods are *singly diagonally implicit Runge–Kutta methods* (SDIRK) which have the property $a_{ij} = 0$ for $i < j$ with all diagonal elements a_{ii} equal ([55]). They satisfy only $C(1)$.

The important property of numerical methods for integration is the order of convergence. We recall that the order of convergence is p if the error, the difference between the exact and numerical solution, is bounded by Ch^p (where C is some constant) on bounded intervals for sufficiently small stepsizes h. Below we cite, after [54], the order of convergence for some classical implicit Runge–Kutta methods. They are valid for ordinary differential equations and for semi–explicit index one system:

$$\begin{aligned} \dot{x}(t) &= f(x(t), y(t)) \\ 0 &= g(x(t), y(t)). \end{aligned}$$

Method	stages	x component	y component
Gauss	$s \begin{cases} \text{odd} \\ \text{even} \end{cases}$	$2s$	$\begin{cases} s+1 \\ s \end{cases}$
Radau IA	s	$2s-1$	s
Radau IIA	s	$2s-1$	$2s-1$
Lobatto IIIA	$s \begin{cases} \text{odd} \\ \text{even} \end{cases}$	$2s-2$	$2s-2$
Lobatto IIIC	s	$2s-2$	$2s-2$
SDIRK	3	3	3 or 2

TABLE 6.1. Runge–Kutta methods: order of convergence.

From Table 6.1 we can deduce that the two–stage Radau IIA method is a good candidate for the integration procedure in our approach to optimal control problems described by large–scale differential–algebraic equations. It has a reasonably high order of convergence: 3. It has the property: $b_i = a_{si}$ which guarantees relatively small number of nonlinear algebraic equations to be solved at every step of an integration procedure. Moreover, it is an $\mathcal{L}$–stable method and thus efficient for stiff equations ([55]). In §6.3 we show that it is possible to provide an efficient procedure for stepsize selection, an important part of the implementation of a numerical integration procedure.

2.2 *Calculation of the Reduced Gradients*

As a result of numerical integration system (1.1) is transformed into a set of nonlinear algebraic equations:

$$\begin{aligned} x(k+1) - x(k) - h(k)\sum_{j=1}^{s} b_j x_j^{'}(k+1) &= 0 \\ F(h^k, x_i^{'}(k+1), x(k) + h(k)\sum_{j=1}^{s} a_{ij} x_j^{'}(k+1), y_i(k+1), u(k)) &= 0, \\ i = 1, \ldots, s & \\ F(h^k, x^{'}(k+1), x(k) + h(k)\sum_{j=1}^{s} b_j x_j^{'}(k+1), y(k+1), u(k)) &= 0, \\ k = 0, \ldots, N_s - 1, \; x(0) = x_0. & \qquad (2.10) \end{aligned}$$

Here, variables $x_i^{'}(k)$, $x^{'}(k)$ correspond to $\dot{x}(t)$, $y_i(k)$ and $y(k)$ to algebraic states, and $x(k)$ to differential states.

If we introduce the notation:

$$X(k) := \left(\left\{ x_i^{'}(k) \right\}_1^s, x^{'}(k), \{y_i(k)\}_1^s, y(k), x(k) \right),$$

then system (2.10) can be considered as a fully implicit discrete time system:

$$\bar{F}(k, X(k+1), X(k), u(k)) = 0, \quad k = 0, \ldots, N_s - 1. \qquad (2.11)$$

Here, $\bar{F}(k, X^+, X, u) : \{0, \ldots, N_s - 1\} \times \mathcal{R}^{(s+1)n+nd} \times \mathcal{R}^{(s+1)n+nd} \times \mathcal{R}^m \to \mathcal{R}^{(s+1)n+nd}$.

Now, our aim is to derive discrete time adjoint equations for system (2.11). If system (1.1) is an index one system and $h(k)$, $k = 0, \ldots, N_s - 1$, are sufficiently small, we can show that the Jacobian matrix of $\bar{F}$ with respect to $X(k+1)$ is nonsingular for all $k = 0, \ldots, N_s - 1$, thus from the Implicit Function Theorem there exists unique function φ such that

$$X(k+1) = \varphi(k, X(k), u(k)), \quad k = 0, \ldots, N_s - 1 \qquad (2.12)$$

and

$$\bar{F}(k,\varphi(k,X(k),u(k)),X(k),u(k)) = 0, \quad k = 0,\dots,N_s-1.$$

Under easily verifiable differentiability assumptions imposed on F the function φ is differentiable with respect to $X(k)$ and $u(k)$, therefore we can write:

$$\bar{F}_{X^+}(k)\varphi_X(k) + \bar{F}_X(k) = 0 \quad \Longrightarrow \quad \varphi_X(k) = -\left[\bar{F}_{X^+}(k)\right]^{-1}\bar{F}_X(k) \tag{2.13}$$

$$\bar{F}_{X^+}(k)\varphi_u(k) + \bar{F}_u(k) = 0 \quad \Longrightarrow \quad \varphi_u(k) = -\left[\bar{F}_{X^+}(k)\right]^{-1}\bar{F}_u(k). \tag{2.14}$$

Here, $\bar{F}_{X^+}(k)$, $\bar{F}_X(k)$, $\bar{F}_u(k)$ are evaluated at $(k, X(k+1), X(k), u(k))$ and $\phi_X(k)$, $\phi_u(k)$ at $(k, X(k), u(k))$.

X is uniquely defined by $\vec{u}$, thus we can write $X^{\vec{u}}$. If we now consider the functional

$$\bar{F}_0(\vec{u}) := \phi(x^{\vec{u}}(N_s)),$$

which is defined by system (2.12), then we are concerned with the adjoint equations as introduced in Chapter 2 (c.f. (2.2.44)):

$$\begin{aligned} p(N_s) &= \phi_X(x^{\vec{u}}(N_s))^T && (2.15)\\ p(k) &= \varphi_X(k)^T p(k+1), \ k = 1,\dots,N_s-1. && (2.16) \end{aligned}$$

Using (2.13) equations (2.16) can be transformed into the equations

$$p(k) \quad = \quad -\bar{F}_X(k)^T\left[\bar{F}_{X^+}(k)\right]^{-T}p(k+1), \ k = 1,\dots,N_s-1$$

and, if we introduce new variables

$$r(k+1) := \left[\bar{F}_{X^+}(k)\right]^{-T}p(k+1) \Rightarrow \bar{F}_{X^+}(k)^T r(k+1) = p(k+1),$$

equations (2.16) can be stated as follows

$$\begin{aligned} \bar{F}_{X^+}(k)^T r(k+1) &= p(k+1), && (2.17)\\ p(k) &= -\bar{F}_X(k)^T r(k+1), \ k = 1,\dots,N_s-1 && (2.18) \end{aligned}$$

with terminal conditions (2.15).

The main computational effort in the calculation of $p(k)$ is associated with solving linear equations (2.17). The Jacobian $\bar{F}_{X^+}(k)$ can be represented as:

$$\bar{F}_{X^+}(k) = \left[\begin{array}{c|c} -h(k)b_1 I_{nd}, \ldots, -h(k)b_s I_{nd}, 0, \ldots, 0 & I_{nd} \\ \hline & \\ J^s(k) & 0 \\ & \end{array} \right], \tag{2.19}$$

where $J^s(k)$ is the system Jacobian corresponding to nonlinear equations in (2.10) and I_{nd} is an identity matrix of dimension nd.

If vector $r(k)$ is composed of $r_1(k)$ and $r_2(k)$ vectors: $r(k) = (r_1(k)^T, r_2(k)^T)^T$ and $p(k)$ is composed of $p_2(k)$ and $p_1(k)$ vectors: $p(k) = (p_2(k)^T, p_1(k)^T)^T$ ($p_1(k)$ is the adjoint variable corresponding to the first equation in (2.12)), then we can easily show that the solution to (2.17) is given by $r_1(k+1) = p_1(k+1)$ and $r_2(k+1)$ which is the solution to the equations

$$J^s(k)^T r_2(k+1) = p_2(k+1) + \begin{bmatrix} b_1 h(k) I_{nd} r_1(k+1) \\ \vdots \\ b_s h(k) I_{nd} r_1(k+1) \\ 0 \\ \vdots \\ 0 \end{bmatrix}.$$

This shows that sparse LU factors of the system Jacobians (c.f. [39]) evaluated during system equation integration can be used to solve adjoint equations.

The adjoint variable p is used to calculate the reduced gradient (according to the convention of (2.3)):

$$\nabla \bar{F}_0(\vec{u})(k) = \varphi_u(k)^T p(k+1), \; k = 0, \ldots, N_s - 1,$$

or, equivalently,

$$\begin{aligned} \nabla \bar{F}_0(\vec{u})(k) &= -\bar{F}_u(k)^T \left[\bar{F}_{X^+}(k)\right]^{-T} p(k+1) \\ &= -\bar{F}_u(k)^T r(k+1), \; k = 0, \ldots, N_s - 1 \end{aligned}$$

and thus can be evaluated with little cost once $r(k+1)$, needed also for the p update, has been calculated.

The reduced gradients for the other functionals $\bar{h}_i^1(\vec{u}) := h_i^1(x^{\vec{u}}(N_s))$, $i \in E$, $\bar{h}_j^2(\vec{u}) := h_j^2(x^{\vec{u}}(N_s))$, $j \in I$ can be obtained in a similar way therefore we only provide the formulas for $\bar{q}(\vec{u})(l) := q(h^l, x^{\vec{u}}(l))$:

$$\begin{aligned}
\nabla \bar{q}(\vec{u})(k) &= 0 \ \forall k \geq l \\
\nabla \bar{q}(\vec{u})(k) &= -\bar{F}_u(k)^T r(k+1), \ k = 0, \dots, l-1 \\
p(l) &= q_X(h^l, x^{\vec{u}}(l))^T \\
\bar{F}_{X^+}(k)^T r(k+1) &= p(k+1) \\
p(k) &= -\bar{F}_X(k)^T r(k+1), \ k = 0, \dots, l-1.
\end{aligned}$$

The reduced gradients can be used to state necessary optimality conditions for the problem $(\mathbf{P}^{\mathbf{N}}_{\mathbf{DAE}})$:

$$\min_{\vec{u}} \phi(x(N_s))$$

subject to the constraints

$$\begin{aligned}
\bar{F}(k, X(k+1), X(k), u(k)) &= 0, \ k = 0, \dots, N_s - 1, \ x(0) = x_0 \\
h_i^1(x(N_s)) &= 0 \ \forall i \in E \\
h_j^2(x(N_s)) &\leq 0 \ \forall j \in I \\
q(h^k, x(k)) &\leq 0, \ k = 1, \dots, N_s - 1 \\
\vec{u} \in \vec{\mathcal{U}} &= \left\{ \vec{u} \in \mathcal{R}^{mN} : \ u(j) \in \Omega, \ j = 0, \dots, N-1 \right\}.
\end{aligned}$$

They are as follows.

Theorem 2.1 *If* $(\bar{X}, \bar{\vec{u}})$ $(\bar{X} = X^{\bar{\vec{u}}})$ *is an optimal pair for the problem* $(\mathbf{P}^{\mathbf{N}}_{\mathbf{DAE}})$ *then there exist numbers* $(\alpha_0, \{\alpha_i^1\}_{i \in E}, \{\alpha_j^2\}_{j \in I}, \{\alpha_q(k)\}_1^{N_s - 1})$ *not all equal to zero such that* $\alpha_q(k) \geq 0$, $k = 1, \dots, N_s - 1$, $\alpha_j^2 \geq 0$, $j \in I$, $\alpha_0 \geq 0$ *and the weak maximum principle condition is satisfied*

$$\min_{\vec{u} \in \vec{\mathcal{U}}} \sum_{k=0}^{N_s - 1} \left[\lambda(k+1)^T \bar{F}_u(k, \bar{X}(k+1), \bar{X}(k), \bar{u}(k)) \left[\bar{u}(k) - u(k) \right] \right] = 0,$$

where λ *satisfy the adjoint equations*

$$\begin{aligned}
&\bar{F}_{X^+}(k-1, \bar{X}(k), \bar{X}(k-1), \bar{u}(k-1))^T \lambda(k) + \\
&\bar{F}_X(k, \bar{X}(k+1), \bar{X}(k), \bar{u}(k))^T \lambda(k+1) = \alpha_q(k) q_X(h^k, \bar{x}(k))^T, \\
&\qquad k = 1, \dots, N_s - 1
\end{aligned}$$

together with the transversality condition

$$\begin{aligned}
&\alpha_0 \phi_X(\bar{x}(N_s))^T + \sum_{i \in E} \alpha_i^1 (h_i^1)_X(\bar{x}(N_s))^T + \sum_{j \in I} \alpha_j^2 (h_j^2)_X(\bar{x}(N_s))^T = \\
&-\bar{F}_{X^+}(N_s - 1, \bar{X}(N_s), \bar{X}(N_s - 1), \bar{u}(N_s - 1))^T \lambda(N_s).
\end{aligned}$$

Moreover,

$$\begin{aligned}
\alpha_j^2 h_j^2(\bar{x}(N_s)) &= 0 \ \forall j \in I \\
\alpha_q(k) q(h^k, \bar{x}(k)) &= 0 \ \forall k \in \{1, \dots, N_s - 1\}.
\end{aligned}$$

Proof.

The proof of the theorem can be obtained by adapting the proof of Theorem 1 in [91] (with obvious changes related to the fact that the 'weak' maximum principle is considered here) and by looking at our formulas for the reduced gradients (defined with the help of the adjoint variables p) where the substitution

$$p(k) = \bar{F}_{X^+}(k-1, \bar{X}(k), \bar{X}(k-1), \bar{u}(k-1))^T \lambda(k), \ k = 1, \ldots, N_s$$

has been made with

$$\begin{aligned} p(N_s) &= -\alpha_0 \phi_X(\bar{x}(N_s))^T - \sum_{i \in E} \alpha_i^1 (h_i^1)_X(\bar{x}(N_s))^T - \\ &\quad \sum_{j \in I} \alpha_j^2 (h_j^2)_X(\bar{x}(N_s))^T. \end{aligned}$$

■

Having the reduced gradients we can apply one of the methods described in Chapters 3–5. For the completeness of the description of our approach we outline in brief the second order algorithm. In order to do that we recall the approximation to ε–active state constraints introduced in Chapter 4 which is tailored here for discrete time systems.

First the intervals $[0, h^{i_0})$, $[h^{i_k}, h^{i_{k+1}})$, i_k, $i_{k+1} \in \mathcal{D}_u$, $k = 0, \ldots, N-1$ (c.f. (2.1)), are divided into n_{sub} equal subintervals:

$$\begin{aligned} [h^{i_{k-1}}, h^{i_k}) &= \bigcup_{j=1}^{n_{sub}} t_j^k, \quad t_i^k \bigcap t_j^k = \emptyset, \\ &\quad i \neq j, \ i, j = 1, \ldots, n_{sub}, \ k = 1, \ldots, N, \\ \text{len}\ (t_j^k) &= \text{len}\ (t_i^k) \ \ \forall i, \ j = 1, \ldots, n_{sub}, \end{aligned}$$

and $h^{i_N} = 1$. Here len (t) is the length of an interval t. Next, from each subinterval t_j^k we choose a point which belongs to the set of ε–active state constraints $R^d_{\varepsilon,\vec{u}}$ defined here as follows

$$R^d_{\varepsilon,\vec{u}} := \left\{ l = 1, \ldots, N_s - 1 : \ \bar{q}(\vec{u})(l) \geq \max_{k=1,\ldots,N_s-1} \bar{q}(\vec{u})(k) - \varepsilon \right\}.$$

To this end we define

$$C_k^j = \left\{ l \in \mathcal{D}_s \bigcap R^d_{\varepsilon,\vec{u}} : \ h^l \in t_j^k \right\}, \ k = 1, \ldots, N, \ j = 1, \ldots, n_{sub}.$$

Then, the approximation to the set $R^d_{\varepsilon,\vec{u}}$ consists from all these points belonging to C_k^j on which $\bar{q}(\vec{u})(\cdot)$ achieves a maximum:

$$j_k^l = \begin{cases} \arg\max \{ \bar{q}(\vec{u})(i) : \ i \in C_k^l \} & \text{if } C_k^l \neq \emptyset \\ \emptyset & \text{if } C_k^l = \emptyset, \end{cases}$$

$$k = 1, \ldots, N, \ l = 1, \ldots, n_{sub}$$

$$A^{n_{sub}} = \bigcup_{k=1}^{N} \bigcup_{l=1}^{n_{sub}} \{j_k^l\}.$$

The optimization algorithm is based on the exact penalty function

$$\bar{F}_c(\vec{u}) = \bar{F}_0(\vec{u}) + cM(\vec{u}).$$

where

$$M(\vec{u}) = \max\left[0, \max_{i\in E}\left|\bar{h}_i^1(\vec{u})\right|, \max_{j\in I}\bar{h}_j^2(\vec{u}), \max_{l\in\mathcal{D}_s}\bar{q}(\vec{u})(l)\right].$$

The direction finding subproblem, $\mathbf{P}_{\mathbf{c,A,H}}^{\mathbf{DAE}}(\vec{\mathbf{u}})$, is as follows

$$\min_{(\vec{d},\beta)}\left[\bar{F}_0(\vec{u}) + \left\langle\nabla\bar{F}_0(\vec{u}), \vec{d}\right\rangle + 1/2(\vec{d})^T H\vec{d} + c\beta\right]$$

$$\begin{aligned}
\text{s. t.} \quad \left|\bar{h}_i^1(\vec{u}) + \left\langle\nabla\bar{h}_i^1(\vec{u}), \vec{d}\right\rangle\right| &\leq \beta \ \forall i \in E \\
\bar{h}_j^2(\vec{u}) + \left\langle\nabla\bar{h}_j^2(\vec{u}), \vec{d}\right\rangle &\leq \beta \ \forall j \in I \\
\bar{q}(\vec{u})(l) + \left\langle\nabla\bar{q}(\vec{u})(l), \vec{d}\right\rangle &\leq \beta \ \forall l \in A \\
d(l) &\in \Omega - u(l), \ l = 0, \ldots, N-1,
\end{aligned}$$

where H is a bounded positive definite matrix and A is an approximation to the set $R_{\varepsilon,\vec{u}}^d$. The problem $\mathbf{P}_{\mathbf{c,A,H}}^{\mathbf{DAE}}(\vec{\mathbf{u}})$ has the unique solution which we denote by $(\bar{\vec{d}}, \bar{\beta})$. Since the solution depends on c. A, H and $\vec{u}$ we may introduce a descent function $\sigma_{c,A,H}^{DAE}(\vec{u})$ and a penalty test function $t_{c,A,H}^{DAE}(\vec{u})$ to be used to test optimality of control $\vec{u}$ and to adjust c, respectively:

$$\begin{aligned}
\sigma_{c,A,H}^{DAE}(\vec{u}) &= \left\langle\nabla\bar{F}_0(\vec{u}), \bar{\vec{d}}\right\rangle + c\left[\bar{\beta} - M(\vec{u})\right], \\
t_{c,A,H}^{DAE}(\vec{u}) &= \sigma_{c,A,H}^{DAE}(\vec{u}) + M(\vec{u})/c.
\end{aligned}$$

Now we can state the algorithm.

Algorithm 5 Fix parameters: $\varepsilon > 0$, γ, $\eta \in (0,1)$, $c^0 > 0$, $\kappa > 1$, $\tau^0 > 0$, $n_0^{sub} = 1$, positive definite matrices H_k, $k = 0, 1, \ldots$

1. Choose the initial control $\vec{u}_0$ such that $u_0(j) \in \Omega$, $j \in \mathcal{D}_u$. Set $l = 0$, $c_{-1} = c^0$, $\tau_0 = \tau^0$, $n_{sub}^0 = n_0^{sub}$.

2. Set $j = 1$. Let $\tau_l^j = \tau_l$, $n_{sub}^j = n_{sub}^l$.

3. Construct $A_l^j = A_l^{n_{sub}^j}$. Let c_l^j be the smallest number chosen from $\{c_{l-1}, \kappa c_{l-1}, \kappa^2 c_{l-1}, \ldots\}$ such that the solution $(\vec{d}_l^j, \beta_l^j)$ to the direction finding subproblem $\mathbf{P}^{\mathbf{DAE}}_{\mathbf{c}_l^j,\mathbf{A}_l^j,\mathbf{H}_l}(\vec{\mathbf{u}}_l)$ satisfies

$$t^{DAE}_{c_l^j,A_l^j,H_l}(\vec{u}_l) \leq 0.$$

If $\sigma^{DAE}_{c_l^j,A_l^j,H_l}(\vec{u}_l) = 0$ then STOP.

Otherwise let α_l^j be the largest number from the finite set $\{\alpha \in \{1, \eta, \eta^2, \ldots\} : \alpha > \tau_l^j\}$ such that

$$\vec{u}_{l+1}^j = \vec{u}_l^j + \alpha_l^j \vec{d}_l^j$$

satisfies the relation

$$\bar{F}_{c_l^j}(\vec{u}_{l+1}^j) - \bar{F}_{c_l^j}(\vec{u}_l) \leq \gamma \alpha_l^j \sigma^{DAE}_{c_l^j,A_l^j,H_l}(\vec{u}_l).$$

If no such α_l^j exists, let $\tau_l^{j+1} = 0.5\tau_l^j$, $n_{sub}^{j+1} = 2n_{sub}^j$, increase j by one and return to Step 3.

Otherwise let $\alpha_l = \alpha_l^j$, $d_l = d_l^j$, $c_l = c_l^j$, $A_l = A_l^j$ and $u_{l+1} = u_{l+1}^j$. Set $n_{sub}^{l+1} = n_{sub}^j$, $\tau_{l+1} = \tau_l$ if $\tau_l^j = \tau_l$ and $n_{sub}^{l+1} = 2n_{sub}^j$, $\tau_{l+1} = 0.5\tau_l$ otherwise. Increase l by one, go to Step 2.

We can show, as in Chapter 5, that *Algorithm 5* is convergent in the sense that

$$\lim_{l \to \infty} \sigma^{DAE}_{c_l,A_l,H_l}(\vec{u}_l) = 0,$$

provided that $\{c_l\}$ is bounded. Because $t^{DAE}_{c_l,A_l,H_l}(\vec{u}_l) \leq 0$ this also implies that

$$\lim_{l \to \infty} M(\vec{u}_l) = 0.$$

We can also show that under some constraint qualification, analogous to $(\mathbf{CQ^N})$, for a given $\vec{u}_l$, A_l, H_l we can always find a finite $c_l \geq c^0$ such that $t^{DAE}_{c_l,A_l,H_l}(\vec{u}_l) \leq 0$. Under the same condition, due to the fact that Ω is compact, we can also prove that $\{c_l\}$ is bounded.

At this point we have to stress that the convergence analysis presented above is valid under the assumption that the discretization for states, $\mathcal{D}_s$, does not depend on $\vec{u}_l$. As it is discussed in the next section stepsizes in the integration procedure are in fact dependent on a current control. Unfortunately to accommodate this feature of the integration procedure in the convergence analysis of the optimization algorithm presented in this chapter would require elaborate considerations on the sensitivity of stepsizes, in a Runge–Kutta method, as a function of controls. It is by no means a trivial problem. This implies, due to our current knowledge of the subject, that we have to apply either the convergence analysis presented in Chapters 3–5, or assume that D_s is fixed.

If we compare our method with the other methods discussed in §6.1, its main drawbacks are:

(i) the method requires a significant amount of computer memory to store the Jacobians $J^s(k)$ (or their sparse LU factors),

(ii) a lack of an efficient stepsize selection procedure for implicit Runge–Kutta methods applied to fully implicit stiff systems (c.f. [55]) although such a procedure has been constructed for non–stiff equations and for stiff systems with separable differential equations (see (3.9)).

3 Implementation of the Implicit Runge–Kutta Method

The essential part of the new approach is an integration procedure. As we have stated in §6.2 at every step of integration we have to solve a set of nonlinear equations (2.10). As we have already pointed out one version of an implicit Runge–Kutta method seems to be very promising for large–scale systems of DAE's. This is the Radau IIA of order 3. Here we recall the coefficients for the method:

$$\begin{array}{c|cc} \frac{1}{3} & \frac{5}{12} & -\frac{1}{12} \\ 1 & \frac{3}{4} & \frac{1}{4} \\ \hline & \frac{3}{4} & \frac{1}{4} \end{array}$$

where the first upper column corresponds to c_i coefficients. the right upper block defines the matrix $\{a_{ij}\}$ and the last right row is the vector $\{b_j\}$.

3.1 Simplified Newton Iterations

At the kth step of the integration procedure we solve the system of equations

$$F^{RK}(w(k+1), x(k), u(k), k) = 0, \tag{3.1}$$

where $w(k)$ is defined as:

$$w(k) := \left(\left\{x_i'(k)\right\}_1^s, \{y_i(k)\}_1^s\right)$$

because $b_i = a_{si}$, $i = 1, \ldots, s$ for the Radau IIA method.

The Newton iteration updates $w_l(k+1)$, $l = 0, 1, \ldots$, first by solving the system of linear equations

$$J_l(k)\Delta w_l(k+1) = -\begin{bmatrix} F_1^l(k) \\ F_2^l(k) \end{bmatrix}, \tag{3.2}$$

where $F_i^l(k) = F(h^k, x'_{i,l}(k+1), x(k) + h(k)\sum_{j=1}^{2} a_{ij}x'_{j,l}(k+1), y_{i,l}(k+1), u(k))$ (Here, we use notation: $w_l(k+1) := (\{x'_{i,l}(k+1)\}_1^s, \{y_{i,l}(k+1)\}_1^s)$.) and the Jacobian of equations (3.1) is as follows

$$J_l(k) = \left[\begin{array}{ccc} F_{\dot{x}}(k)_1^l + a_{11}h(k)F_x(k)_1^l & a_{12}h(k)F_x(k)_1^l & F_y(k)_1^l \\ a_{21}h(k)F_x(k)_2^l & F_{\dot{x}}(k)_2^l + a_{22}h(k)F_x(k)_2^l & 0 \end{array}\right.$$

$$\left.\begin{array}{cc} \rightarrow & 0 \\ \rightarrow & F_y(k)_2^l \end{array}\right]$$

where, for example,

$$\begin{aligned} F_{\dot{x}}(k)_1^l &= F_{\dot{x}}(h^k, x'_{1,l}(k+1), x(k) + \\ &\quad h(k)\sum_{j=1}^{2} a_{1j}x'_{j,l}(k+1), y_{1,l}(k+1), u(k)). \end{aligned}$$

Having the solution to equations (3.2) $w(k+1)$ is updated by substituting $w_{l+1}(k+1) = w_l(k+1) + \Delta w_l(k+1)$. The initial guess for $w_0(k+1)$ can be $w(k)$ although better choices are possible (c.f. [55]). In order to reduce the number of Jacobians evaluations we assume that

$$F_{\dot{x}}(k)_1^l = F_{\dot{x}}(h^k, x'(k), x(k), y(k), u(k)),\ l = 0, 1, \ldots, \tag{3.3}$$

and the same for the other involved Jacobians (unless the progress in solving equations (3.2) is not sufficient then $w_l(k+1)$ is used in (3.3) instead of $w(k)$). The iteration (3.2) modified in this way we call the simplified Newton iteration.

3.2 Stopping Criterion for the Newton Method

If the Newton iterations converge we have

$$\|\Delta w_{l+1}(k+1)\| \le t\,\|\Delta w_l(k+1)\|,\quad t < 1,$$

and the well–known result (e.g., [113])

$$\|w(k+1) - w_{l+1}(k)\| \le \frac{t}{1-t}\,\|\Delta w_l(k+1)\|, \tag{3.4}$$

where $w(k+1)$ is a solution to (3.1), can be used to stop Newton's iterations because (3.4) implies that $w_{l+1}(k)$ is a good approximation to a solution $w(k+1)$.

In reality t is not a constant and should be estimated, for example, by

$$t_l = \frac{\|\Delta w_l(k+1)\|}{\|\Delta w_{l-1}(k+1)\|}, \ l \geq 1$$

and we stop computations when

$$\frac{t_l}{1-t_l}\|\Delta w_l(k+1)\| \leq \kappa\varepsilon, \tag{3.5}$$

where $\kappa \in (0.1, 1)$ and $\varepsilon > 0$ is close to a local discretization error although as it is explained later in the section there are situations when this value should be much smaller than this error. This strategy can only be applied after at least two iterations.

We allow the maximum number of iterations l_{max} to be performed and if during these iterations (3.5) is not satisfied we change the stepsize $h(k) := h(k)/2$ and start solving a new set of nonlinear algebraic equations. We also stop computations and change the stepsize, in the same way, if at some iteration $t_l > 1$.

3.3 Stepsize Selection

Because the number of steps in integration can have a strong effect on the computational time of adjoint equations, the stepsize selection procedure is a very important part of our implementation of the Runge–Kutta method. Therefore, we are interested in keeping this number as small as possible to reduce not only the integration time. Furthermore, our stepsize selection procedure should be suitable for stiff equations which are very likely to arise in optimal control problems where the basic problem concerns an open loop stabilization.

As we have already mentioned we do not expect to use high order integrations schemes to optimize large–scale problems. In order to keep the number of nonlinear equations in (3.1) as small as possible we prefer the 2–stage Radau IIA scheme. Therefore, we concentrate on the local error estimation for this particular method. The fundamental equations we consider are the following

$$x(k+1) = x(k) + h(k)\sum_{j=1}^{2} b_j x_j^{'}(k+1).$$

Because the 2–stage Radau IIA method is an optimal scheme, i.e., it cannot be embedded into a higher–order integration scheme with the same number of stages, we have to consider lower–order formulas which could be derived from the set of nonlinear equations we solve at every stage.

One such scheme could have the fundamental equations

$$x^{lo}(k+1) = x(k) + h(k)\left[b_0^{lo}x'(k) + \sum_{j=1}^{2} b_j^{lo}x_j'(k+1)\right].$$

The coefficients $\{b_j^{lo}\}$ should be chosen in such a way to guarantee

$$\begin{aligned} x^{lo}(k+1) - x(k+1) &= h(k)\left[b_0^{lo}x'(k) + \sum_{j=1}^{2}[b_j^{lo} - b_j]x'(k+1)\right] \\ &\approx O(h(k)^2). \end{aligned} \quad (3.6)$$

Having in mind the basic principles which are used to calculate $\{a_{ij}\}$, $\{b_j\}$ for a given Runge–Kutta method the requirement (3.6) is met if $\{b_j^{lo}\}$ satisfy the equations[3]

$$\begin{aligned} b_0^{lo} + \sum_{j=1}^{2} b_j^{lo} &= 1 \\ \sum_{j=1}^{2} b_j^{lo} c_j &= \frac{1}{2}. \end{aligned} \quad (3.7)$$

If we fix b_0^{lo}, the other values b_j^{lo} can be calculated from (3.7).

As Hairer *et al* ([55]) pointed out the error estimate based on (3.6)–(3.7) is not suitable for stiff equations. If we consider the equation $\dot{x}(t) = \lambda x(t)$ and $h(k)\lambda \to \infty$ the difference in (3.6) will behave like $[x^{lo}(k+1) - x(k+1)]/x(k) \approx b_0^{lo}h(k)\lambda$ (this can be shown if we notice that $x_j'(k+1) \to 0$, $j = 1, 2$ when $h(k)\lambda \to \infty$) and thus will be unbounded. If we consider again the same simple equation then we can find that for the error defined by[4]

$$\varepsilon^l = \left[I - h(k)b_0^{lo}J_k\right]^{-1}\left[x^{lo}(k+1) - x(k+1)\right], \quad (3.8)$$

where J_k is a Jacobian of f in the equation $\dot{x}(t) = f(x(t))$, for $h(k)$ close to zero we have $\varepsilon^l = O(h(k)^2)$ but for $h(k)\lambda \to \infty$, $\varepsilon^l/x(k) \to -1$ which is much better than for the formula defined by (3.6)–(3.7).

Now the error formula (3.8) is translated to a system of differential–algebraic equations

$$\begin{aligned} f(t, \dot{x}(t), x(t), y(t), u(t)) &= 0 \\ g(t, x(t), y(t), u(t)) &= 0 \end{aligned} \quad (3.9)$$

[3] These equations can be derived in the same way as equalities $B(p)$ (see (2.7)), rather elaborate details of the derivation are omitted.

[4] an idea allegedly attributed to Shampine according to [55]

where

$$\det\left[f_{\dot{x}}(t,\dot{x}(t),x(t),y(t),u(t))\right]\neq 0,\quad \det\left[g_y(t,x(t),y(t),u(t))\right]\neq 0$$

for all $\dot{x}(t)$, $x(t)$, $y(t)$ and t (the assumption which implies that the index of system (3.9) is one). Assuming that all data are sufficiently smooth we can use the Implicit Function Theorem to state

$$\begin{aligned}\dot{x}(t) &= \varphi_1(t,x(t))\\ y(t) &= \varphi_2(t,x(t))\end{aligned}$$

and

$$\begin{aligned}(\varphi_1)_x(t) &= -\left[f_{\dot{x}}(t)\right]^{-1}\left[f_x(t)+f_y(t)(\varphi_2)_x(t)\right]\\ (\varphi_2)_x(t) &= -\left[g_y(t)\right]^{-1}g_x(t)\end{aligned}$$

for some functions φ_1,φ_2. Here, $f_{\dot{x}}(t)$, $f_x(t)$, $f_y(t)$, $g_x(t)$, $g_y(t)$ are evaluated at $(t,\dot{x}(t),x(t),y(t),u(t))$ and $(\varphi_1)_x(t)$, $(\varphi_2)_x(t)$ at $(t,x(t))$.

Therefore, if the system equations are separable (see (3.9)) and conditions (3.7) are satisfied then the generalization of (3.8) can be stated as

$$\begin{aligned}&\left[f_{\dot{x}}(k)+h(k)b_0^{lo}\left[f_x(k)-f_y(k)\left[g_y(k)\right]^{-1}g_x(k)\right]\right]\varepsilon^l\\ &\quad = f_{\dot{x}}(k)\left[x^{lo}(k+1)-x(k+1)\right].\end{aligned}\tag{3.10}$$

Here, for example, $f_{\dot{x}}(k)=f_{\dot{x}}(h^k,x'(k+1),x(k+1),y(k+1),u(k)),\ g_y(k)=g_y(h^k,x(k+1),y(k+1),u(k))$.

The error estimates should also satisfy the equations

$$g(h^k,x(k+1)+\varepsilon_x^l,y(k+1)+\varepsilon_y^l)=0,$$

thus

$$g(k)+g_x(k)\varepsilon_x^l+g_y(k)\varepsilon_y^l=0,$$

where $g(k)$ are residuals for actually calculated states: $g(k)=g(h^k,x(k+1),y(k+1),u(k))$, and

$$\varepsilon_y^l=-\left[g_y(k)\right]^{-1}\left[g_x(k)\varepsilon_x^l+g(k)\right].$$

Now, if we assume that $\varepsilon_x^l=\varepsilon^l$ and ε_y^l corresponds to this error, we can rewrite equation (3.10):

$$\begin{aligned}\left[f_{\dot{x}}(k)+h(k)b_0^{lo}f_x(k)\right]\varepsilon^l &= f_{\dot{x}}(k)\left[x^{lo}(k+1)-x(k+1)\right]-\\ &\quad h(k)b_0^{lo}f_y(k)\left[\varepsilon_y^l+\left[g_y(k)\right]^{-1}g(k)\right].\end{aligned}$$

Therefore, if we assume that $g(k) \approx 0$ (as we did when we proposed equations (3.6)) and that the influence of ε_y^l on ε^l is neglected then we come to the error estimate

$$\left[f_{\dot{x}}(k) + h(k)b_0^{lo} f_x(k)\right] \varepsilon^l = f_{\dot{x}}(k)\left[x^{lo}(k+1) - x(k+1)\right]. \quad (3.11)$$

Notice that the complete equations we could use for ε^l are much more complex:

$$\begin{bmatrix} f_{\dot{x}}(k) + h(k)b_0^{lo} f_x(k) & h(k)b_0^{lo} f_y(k) \\ g_x(k) & g_y(k) \end{bmatrix} \begin{bmatrix} \varepsilon^l \\ \varepsilon_y^l \end{bmatrix} = \begin{bmatrix} f_{\dot{x}}(k)\left[x^{lo}(k+1) - x(k+1)\right] \\ -g(k) \end{bmatrix}. \quad (3.12)$$

Note that these equations give error estimates for both x and y.

The error estimate ε^l is used to determine the stepsize according to the rule:

$$h_{new} = \tau h_{old} \left(\frac{\varepsilon}{\|\varepsilon^l\|}\right)^{0.5}, \quad (3.13)$$

where, as in (3.5), ε is a local discretization error and $\tau = 0.8$.

Our numerical experience with the error equation (3.11) shows that the error estimate (3.6) can be improved by a factor of two for bigger stepsizes. As a result computational time can be significantly reduced (due to fewer steps in integration) even though an extra factorization of the matrix in the left–hand side of equation (3.11) is needed to find ε^l. The values of errors calculated according to different formulas, when (3.6) was used to determine steps of integrations, are presented in Figs 6.1–6.3. These experiments correspond to the problem stated in Example 2 ($N = 50$ in this experiment), Example 3 ($N = 50$) and in Example 4 ($N = 50$). Note that the scalar $\varepsilon/\|\varepsilon^l\|$ is shown on the figures.

They show that the error calculated according to (3.11) is a good approximation to the error formula (3.12) we should use for stiff equations. The plausible explanation is that in order to calculate ε^l satisfying (3.12) we have to modify the right–hand side of (3.11) by the term $h(k)b_0^{lo} f_y(k)\varepsilon_y^l$. However, the small ε_y^l is multiplied by $h(k)$, which for our normalized horizon $[0, 1]$ and the discretization for control variables typically at least equal to 50, is less than 0.02. According to our experience using (3.11) instead of (3.12) does not affect significantly calculated solutions to control problems.

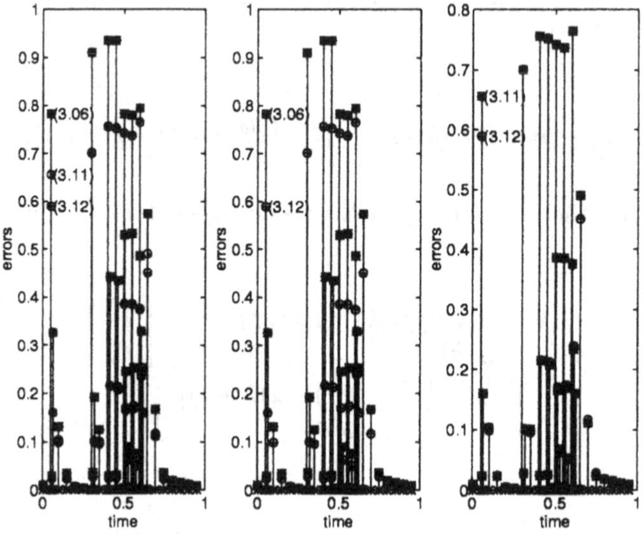

FIGURE 6.1. Error estimates, Example 2.

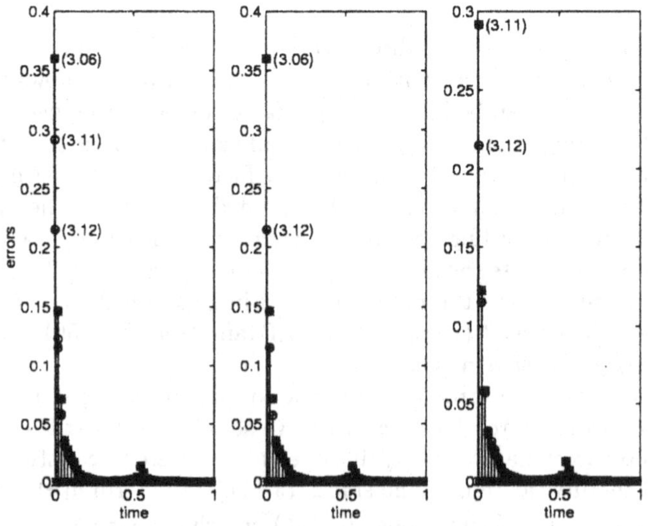

FIGURE 6.2. Error estimates, Example 3.

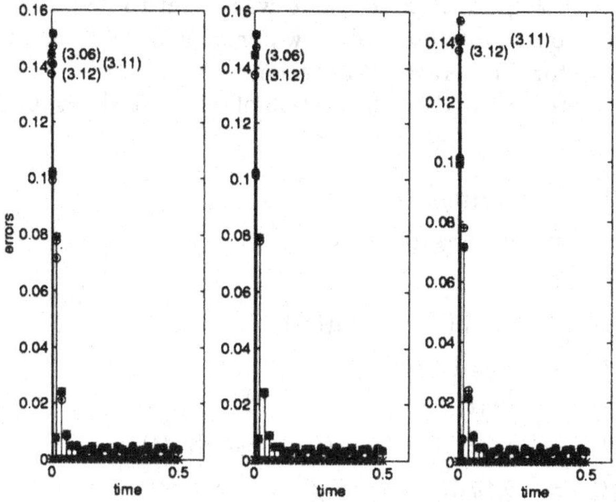

FIGURE 6.3. Error estimates, Example 4.

4 Numerical Experiments

The Radau IIA method was implemented with the help of the MA28 numerical package for solving large–scale sparse systems of linear equations ([58]). We found the package specially well–suited to our application. At the first step MA28A found a sparse LU decomposition of the Jacobian $J^s(k)$ using the pivotal strategy to compromise between maintaining sparseness and controlling loss of accuracy through roundoff errors. Then MA28B used the pivotal sequence determined by MA28A to factorize, at much lower cost, the subsequent Jacobians $J^s(k)$. We applied the BFGS updating scheme with Powell's modifications ((4.80)–(4.84)) to generate matrices H_l used in direction finding subproblems. Numerical experiments were performed on a SUN Sparc Classic workstation with 32 Mb of main memory.

Several optimal control problems have been solved using *Algorithm 5*. For illustration we cite only some of them. The full description of these problems is given in [106] in the form of gPROMS files ([8]) (with the exception of the problem discussed in Example 5). The gPROMS simulation program was used to define systems dynamics, then FORTRAN files describing residuals and sparse Jacobians were translated to our optimization package.

Example 1 (Batch reactor). We consider an exothermic reaction scheme, A + B → C + D, taking place in a jacketed batch reactor. Here, the objective was to find the best cooling water control policy so as to maximize the yield of component C in the presence of constraints on the reactor temperature. The system equations consisted of 10 (n=10) differential–algebraic equations, describing the mass and energy balances and there were 4 (nd=4) differential states. The number of nonzero elements in the Jacobian of F with respect to $\dot{x}$, x and y was equal to 38 (nz=38). The problem had box constraints on cooling water control and one state constraint for the reactor temperature limitation.

The process was represented by the system of differential–algebraic equations:

$$
\begin{aligned}
\dot{x}_i(t) &= c_i y_4(t) y_5(t), \ i = 1, 2 \\
\dot{x}_3(t) &= d y_4(t) y_5(t) - y_8(t) \\
\dot{x}_4(t) &= u(t) \\
y_6(t) &= a \exp(-e/(f y_{10}(t))) \\
y_4(t) &= y_6(t)(y_7(t))^2 \\
y_8(t) &= g u(t) \\
x_3(t) &= x_1(t)[y_1(t) + y_2(t)] + x_2(t)[y_2(t) + y_3(t)] \\
y_i(t) &= b_i(y_9(t) - h), \ i = 1, \ldots, 3 \\
x_1(t) &= y_5(t) y_7(t) \\
y_5(t) &= x_1(t)/\rho_1 + x_1(t)/\rho_2 + 2x_2(t)/\rho_3 \\
y_{10}(t) &= 2x_2(t) - x_4(t), \ \ t \in [0, t_f],
\end{aligned}
$$

where $a = 7.5 \cdot 10^{-4}$, $b_1 = 150$, $b_2 = 175$, $b_3 = 200$, $c_1 = -1$, $c_2 = 1$, $d = 6.0 \cdot 10^4$, $e = 1.9 \cdot 10^4$, $f = 8.314$, $g = 1.68 \cdot 10^5$, $\rho_1 = 1.0 \cdot 10^4$, $\rho_2 = 8.0 \cdot 10^3$, $\rho_3 = 1.1 \cdot 10^4$ and $h = 296$. The initial conditions were: $x_1(0) = 5000$, $x_2(0) = 0.0$, $x_4(0) = 0.0$, $y_{10}(0) = 300$; the initial control: $u_0(t) = 0.0$, $t \in [0, t_f]$ and the horizon was: $t_f = 1000$.

The objective function and constraints complete the description of the problem:

$$\max_u y_{10}(t_f)$$

subject to

$$
\begin{aligned}
y_9(t) - 400 &\le 0, \ \ t \in [0, t_f] \\
0.1 \le u(t) &\le 3, \ \ t \in [0, t_f]
\end{aligned}
$$

and

$$y_9(t_f) = 315.$$

After discretization we had 50 (N=50) control parameters to be optimized. The summary of the numerical experiments is given in Table 6.2. The calculated control is presented in Figure 6.4.

The variants 1 and 2 in Table 6.2 correspond to the cases when factorized Jacobians created during integration were stored and then used in evaluating adjoint equations. In the variant 1 all state trajectories and time

Variant	1	2	3	4
Systems equations	29.42	28.9	30.02	29.82
Adjoint equations	7.47	6.39	19.64	17.92
SQP algorithm	6.4	5.98	6.41	6.17
Total	43.29	41.27	56.07	53.91

TABLE 6.2. Example 1, CPU time (sec).

derivatives of differential states from all stages of integration were stored to be used in the integration of adjoint equations. In variant 2 only those states which were present in state constraints, furthermore sparse Jacobians $\bar{F}_u(k)$ were evaluated and stored during the integrations of systems equations.

The difference between variants 3–4 and 1–2 is that in the former instead of storing factorized Jacobians the Jacobians themselves were stored.

Example 2. This is a problem of stabilizing, in an open loop, the CO_2 absorption/stripping process shown in Figure 6.5. The system considered is a gas sweetening unit which removes carbon dioxide from a nitrogen gas stream by means of a monoethanolamine (MEA) solution. The carbon dioxide is absorbed by the MEA solution in a packed absorption column and is removed from the solution by a stream stripping column. The objective is to keep the concentration of carbon dioxide at some operating point in the face of disturbances in the feed composition while at the same time utilizing minimum resources. The control variables employed were the steam flow through the stripping column and the MEA circulation rate. The problem had 188 differential–algebraic equations (n=188) among them were 27 differential equations (nd=27). The number of nonzero elements in the Jacobian of F was equal to 660 (nz=660). In addition to two terminal equality constraints we had mixed control only constraints at every instant of time. The number of controls was equal to 2 and the discretization for controls was 100 (N=100) thus in total we had 200 decision variables. The results are summarized in Table 6.3, the calculated controls are shown in Figure 6.6. The 'steady' behaviour of optimal solutions is reflected in low

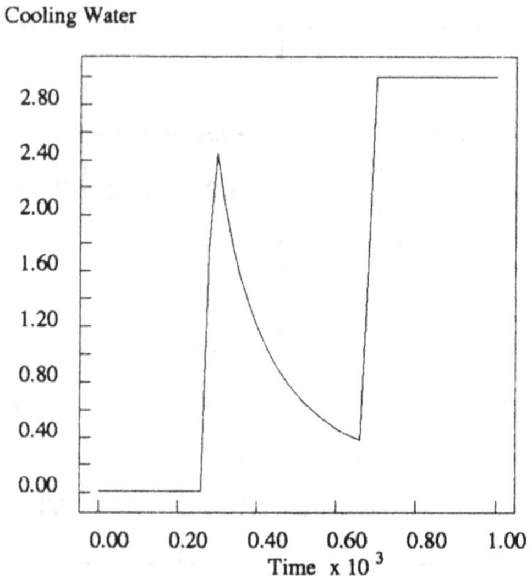

FIGURE 6.4. Example 1, control profile.

computing time for this example.

Example 3 (Distillation column). The third and fourth examples are related to the design of an optimal controller for a ternary distillation column (n-hexane, heptane and toluene), so as to minimize the total annualized operating cost under specified variability ([78], see also Figure 6.7). The process model describing the distillation column is given by a set of differential–algebraic equations for mass, composition, energy and equilibrium relationships on a tray–by–tray basis for all the column compartments, the rectifying, stripping, condenser, reflux–drum, column base, reboiler and bottom tray sections, respectively. The detailed mathematical model is described in [78]. Here, the distillation column we consider consisted of 23 trays given by 768 differential–algebraic equations (n=768) and 103 differential states (nd=103). The number of nonzero elements in the Jacobian of F was 2659 (nz=2659). There were 3 controls which were discretized on 50 subintervals (N=50). The problem had only one state constraint corresponding to the requirement on the output composition. The results are presented in Table 6.4.

The problem was also solved by using the full discretization approach described in [117] and the modelling language GAMS ([23]) with the NLP solver CONOPT ([38]). Here, 3 orthogonal collocation points on 10 finite elements were used. The discretization of control variables was thus 5 times lower than in the run of our method yet the CPU time of 29 500 sec. was significantly higher. Notice that the program CONOPT was implemented

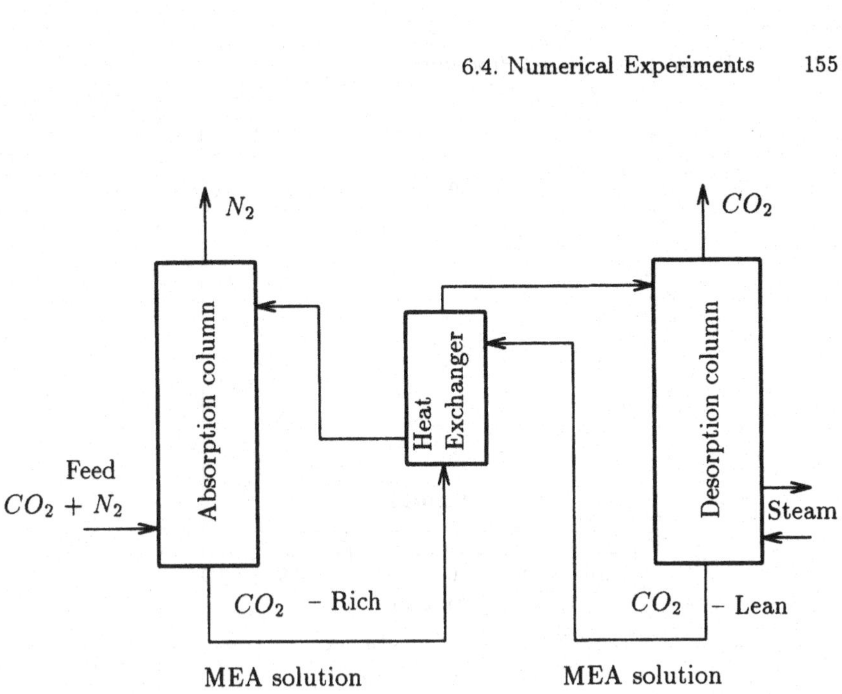

FIGURE 6.5. Example 2, CO_2 absorption/stripping process.

Variant	1	2	3	4
Systems equations	188.20	181.19	179.69	172.84
Adjoint equations	36.73	32.25	139.12	131.38
SQP algorithm	18.56	18.84	18.84	18.85
Total	240.0	230.6	334.7	322.2

TABLE 6.3. Example 2, CPU time (sec).

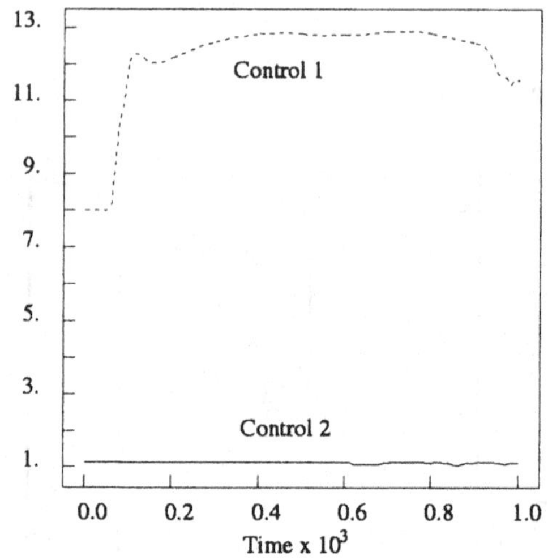

FIGURE 6.6. Example 2, control profiles.

using sparse matrix techniques. (This remark applies also to the program described in [119] and used in comparisons later in this section. The program in [119] is based on sensitivity equations (1.10) and the implementation of BDF method.) This very approximate comparison with the full discretization approach indicates the efficiency of our method.

We have to stress that the comparison of our method with the other approaches considered here must be treated very cautiously. Performance of these methods depends on several parameters which, in many cases, cannot be changed by a user—those we could change we set to values as close as possible to the values of the corresponding parameters in our program. However, we think that CPU times recorded by CONOPT and the program based on the sensitivity equations adequately reflect their performance on large–scale problems.

Example 4 (Distillation column). The problem is similar to that described in the previous example, the difference being column design, control structure and product specifications. In this case the ternary distillation column consisted of 21 trays given by 702 differential–algebraic equations (n=702), 96 differential states (nd=96) and two controls (reflux ratio and reboiler duty) which were defined on 50 subintervals (N=50). The number of nonzero elements in the Jacobian of F was 2409 (nz=2409). Moreover the problem had 4 state constraints to ensure that the distillate and bottom specifications were met in spite of a feed flowrate disturbance. The results obtained using our approach are presented in Table 6.5 and in Figure 6.8.

The results for variants 3 and 4 emphasize that every detail of a compu-

Variant	1	2	3	4
Systems equations	3551.05	3431.61	3361.07	3321.05
Adjoint equations	160.20	134.41	2280.55	2203.67
SQP algorithm	14.2	13.71	12.68	12.50
Total	3700.0	3566.7	5646.2	5533.0

TABLE 6.4. Example 3, CPU time (sec).

Variant	1	2	3	4
Systems equations	2126.99	2160.94	2048.87	2093.83
Adjoint equations	1758.72	1761.52	26614.47	24110.88
SQP algorithm	8.97	8.26	6.81	6.93
Total	3370.0	3471.5	28666.0	26208.3

TABLE 6.5. Example 4, CPU time (sec).

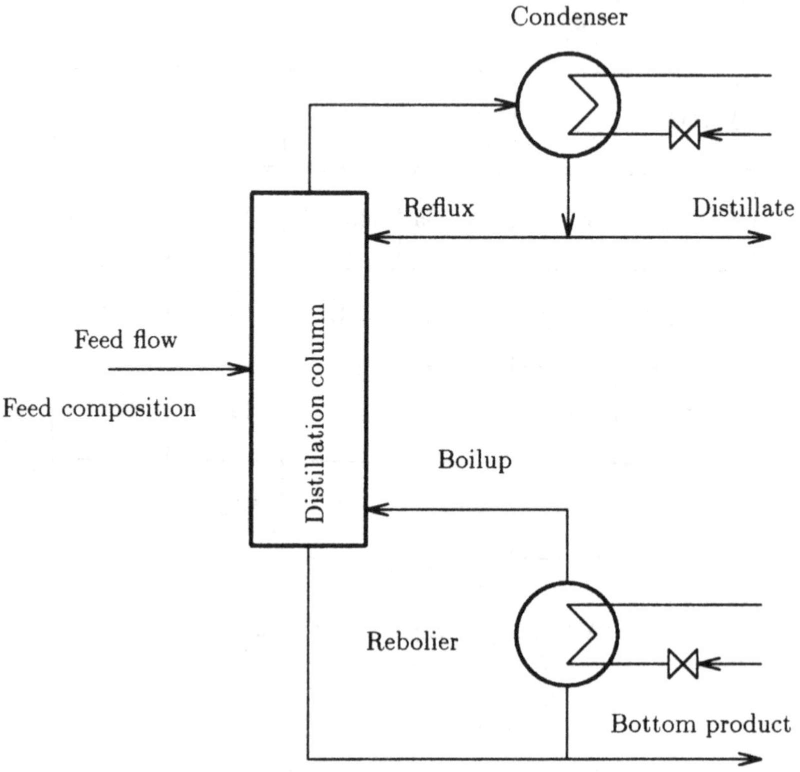

FIGURE 6.7. Examples 3–4, distillation column.

tational procedure for optimal control of the large system of differential–algebraic equations has to be thoroughly worked out, if we want to avoid excessive computational time. The significant difference for variants 1–2 and 3–4, for this example, can be explained by very irregular behaviour of the solutions. As a result of this Jacobians $J^s(k)$ (c.f. (2.19)) changed from one step of the integration to another.

The algorithm based on sensitivity equations ([119]) found controls shown in Figure 6.9[5] (discretization for controls was $N = 10$) and needed a factor of two and a half times more CPU time than *Algorithm 5*. In fact the program described in [119] stopped when a feasible point for constraints (1.2) was found. This was due to the fact that in [119] state constraints (1.2) are

[5]The figure also illustrates that too crude piecewise constant approximations provide very little information about control functions.

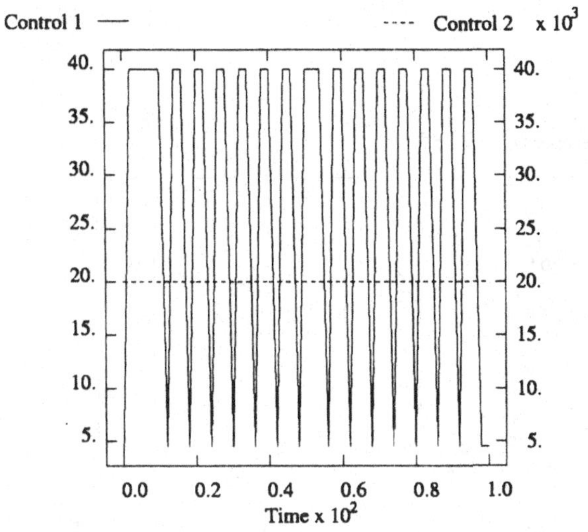

FIGURE 6.8. Example 4, optimal control profiles.

substituted by the equality constraint

$$\begin{aligned} \bar{h}(\vec{u}) &= h(x^{u^N}(1)) = \int_0^1 \left[\max\left[0, q(t, x^{u^N}(t))\right]\right]^2 dt \\ &= x_{n+1}^{u^N}(1) = 0. \end{aligned} \tag{4.1}$$

Once (4.1) is satisfied $\nabla \bar{h}(\vec{u}) = 0$ and the program stops.

The application of the full discretization approach implemented as discussed before (the number of finite elements equal to 10) resulted in a CPU time of 36 720 sec.

The performance of the SQP algorithm for all four examples is reported in Table 6.6. The constraints for all examples were satisfied with the accuracy 10^{-5}, the stopping criterion was $\sigma^{DAE}_{c_k,A_k,H_k}(u_k) \geq -10^{-3}$.

Example 5 We consider the optimal control problem of the large–scale industrial process treated in [120] as part of a collaborative project between Imperial Chemical Industries (ICI) and the Centre for Process Systems Engineering at Imperial College.[6] It consists of two adiabatic plug flow reactors (PFRs) in series, where a multicomponent feed undergoes exothermic reactions (Fig. 6.10). Process constraints apply to both inlet and output temperatures, as well as to the product concentration of one of the components. The control purpose is to minimize the production rate

[6]the model of the process developed at ICI was available to us and we kindly acknowledge it here.

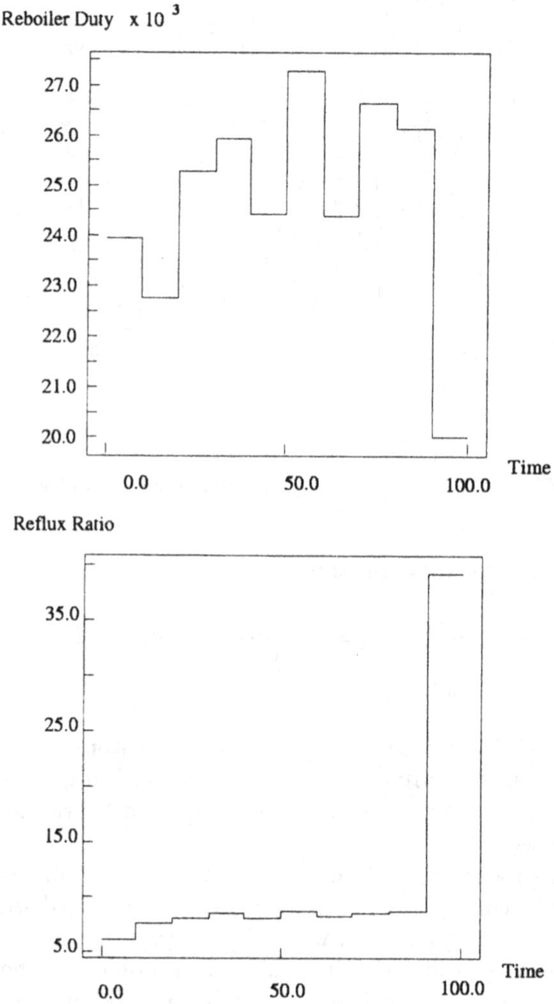

FIGURE 6.9. Example 4, control profiles—sensitivity eqns. approach.

Example	1	2	3	4
Number of iterations	15	11	51	15
Number of systems equations evaluations	62	32	52	39
Number of adjoint equations evaluations	337	33	104	480
$\sigma^{DAE}_{c_l,A_l,H_l}(\vec{u}_l)$ at final iteration	$-1.5 \cdot 10^{-6}$	$-7.5 \cdot 10^{-4}$	$-3.1 \cdot 10^{-4}$	$-4.1 \cdot 10^{-6}$

TABLE 6.6. Performance of SQP algorithm.

of a byproduct. Process constraints and critical disturbance scenario must be considered in order to guarantee feasibility and low cost of steady state operations. This gives rise to a dynamic optimization problem in the framework of the so-called *steady-state backoff problem* ([4]).

The process model was set up in gPROMS ([8]), exhibiting $n = 2678$ differential-algebraic equations and $nd = 283$ differential equations. The Jacobian $J^s(k)$ had $nz = 8261$ nonzero elements.

The inlet temperatures T_{in1} and T_{in2} are the control variables, while the exit temperatures T_{out1} and T_{out2} are controlled variables. Two disturbances show a dominant effect: feed flow and the concentration of one of the reactants. Results related to the controllability analysis of the process are fully described in [120]. We present the solution of the steady-state backoff problem.

The steady-state backoff problem can be posed as follows. A process is in steady-state operation without or with minimum control action at a point determined by optimising the process defined by some nominal values of uncertain parameters. Feasible operation cannot be guaranteed when the values of the uncertain parameters change. Then, the goal is to obtain another operating point (if it exists) such that: a) it is as close as possible to the previous operating point and b) there exist controls which bring the process, perturbed by some critical disturbances, back to this new operating point. In other words, the new point remains feasible under the critical disturbances and is as close as possible to the steady-state economics. This point is called the *backoff point* ([4]).

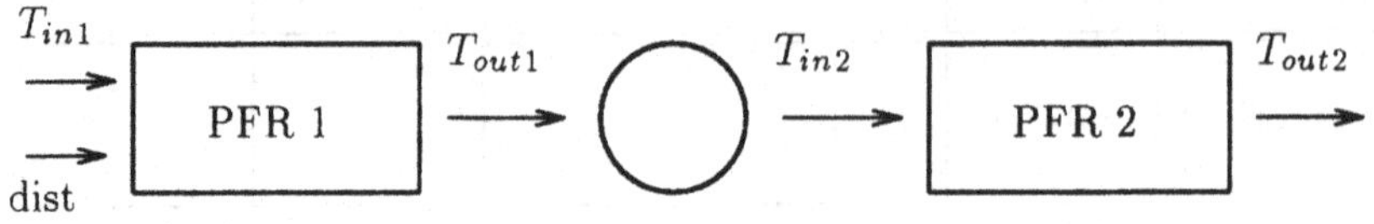

FIGURE 6.10. Example 5, PFD process.

The minimisation of the production rate of a byproduct (called fB) at the nominal conditions was used as the economic objective. In [120] a critical disturbance scenario for T_{out1} was used to estimate T_{in1}, T_{in2} which guarantee feasibility of T_{out1}, T_{out2} and low cost of steady state operations.

Since the formulation of the problem contains constraints on initial values of differential states, we have to describe the modifications to the evaluation of adjoint equations which do not increase the dimension of the optimization problem.

We have the following additional equations to define the initial conditions for system (2.11):

$$G(\dot{x}(0), x(0), y(0), u(0)) = 0. \tag{4.2}$$

If $x(0)$ are given, then $\dot{x}(0)$ and $y(0)$ are calculated (for a given $u(0)$) using equations (4.2), and standard adjoint equations can be applied to any functional in the problem ($\mathbf{P}^{\mathbf{N}}_{\mathbf{DAE}}$).

However, if $\dot{x}(0)$ is fixed (or its part together with the part of $y(0)$) then $x(0)$ is, through (4.2), a function of $u(0)$. This fact must be reflected in the redefined adjoint equations. Notice that we could avoid redefining adjoint equations by introducing new optimization parameters corresponding to varying $x(0)$ at the cost of significantly complicating the optimization problem.

If we consider the additional equations (4.2) which determine $X(0)$, we have to modify the Lagrangian by adding the term

$$\lambda_{in}^T G(X(0), u(0)),$$

where λ_{in} are the Lagrange multipliers corresponding to equations (4.2) (we use concise notation for $\dot{x}(0)$, $x(0)$, $y(0)$ introduced earlier)—compare with §2.2.

The new Lagrangian will look as follows

$$\lambda_{in}^T G(X(0), u(0)) + \lambda(1)^T \bar{F}(0, X(1), X(0), u(0)) + \dots \tag{4.3}$$

if only the adjoint variable corresponding to the first time instant, $\lambda(1)$, is listed.

Now, we would like to remind that in order to calculate reduced gradients we take partial derivatives of (4.3), with resepct to $X(k)$, $k = 0, \ldots, N-1$ and equal them to zero (see §2.2). This results in the equation

$$\lambda_{in}^T G_X(X(0), u(0)) + \lambda(1)^T \bar{F}_{X^+}(0, X(1), X(0), u(0)) = 0. \tag{4.4}$$

Since G_X is nonsingular (in fact we assume that—the system has index one) and $\lambda(1)$ is given as the solution to standard adjoint equations which do not take into account equations (4.2), λ_{in} can be calculated from equations (4.4).

λ_{in} is then used in the gradient equations

$$\lambda_{in}^T G_u(X(0), u(0)) + \lambda(1)^T \bar{F}_u(0, X(1), X(0), u(0)) + \ldots = 0,$$

which lead directly to the reduced gradient. Observe that λ_{in} is uniquely defined for each set of adjoint equations required for each functional in the problem.

The extra cost associated with this modification of adjoint equations is negligible in comparison to the cost of evaluation of system and adjoint equations. Note also that the factorization of the matrix G_X is available from the initialization step of system equations integration.

In [120], the nonlinear optimal control analysis was carried out for one critical disturbance scenario. The restricted step disturbance set appearing in the output variable T_{out1} was used. The optimal value of the objective function (value of fB variable) obtained by the program described in [119] was 0.056 ([120]) with more than 100 iterations performed, using the discretization $N = 42$.

Some state constraints were outside their bounds at the beginning of horizon when a sudden drop of about 5% occurs in the disturbance. For the whole horizon, the open–loop optimal control law imposed a very hard control action on both reactors. Considerable CPU time (counted in days on SUN SPARCstation Model 10) was necessary to obtain the solution. This was a consequence, on one hand, of the path constraint representation (4.1) and, on the other hand, of the evaluation of gradients by sensitivity equations. Howvever, we have to admit that, the cost of the computation could be reduced by making some changes in the implementation of sensitivity equations available code which would enable to decrease the accuracy in the integration procedure.

We solved the same problem using the method introduced in this chapter (only the first variant of the implementation of the method was run on the problem), considering the discretization $N = 50$. In Figure 6.11, the optimal control for the first reactor is shown. The corresponding output is compared with that from the sensitivity equations based approach in Figure 6.12. No constraints were violated and the control variables were smooth. The solution was obtained in about 20 minutes of CPU time, on a SunSPRAC ultra1 workstation (the program needed about 60 Mbytes of main memory), after performing 14 iterations, with absolute and relative accuracies in integration set to 10^{-4}—[33]. The backoff point was close to that reported in [120] (objective function: 0.0596—the lower value of the objective function for the control calculated by the program described in [119] can be explained by the fact that constraints were violated at this point). Regarding to the second reactor, the solution obtained by our method reflects the qualitative analysis reported in [120]. The input was at the lower bound and the output was constant and close to its lower bound.

We also solved the problem with the discretization of controls: $N = 100$. The solution was obtained in 30 iterations in CPU time 1.86 hrs (accuracies for integration were set at 10^{-3}) with $\sigma^{DAE}_{c_l,A_l,H_l}(u_l)$ at the last iteration (as for the run with $N = 50$) satisfying $\sigma^{DAE}_{c_l,A_l.H_l}(u_l) \geq -10^{-3}$.

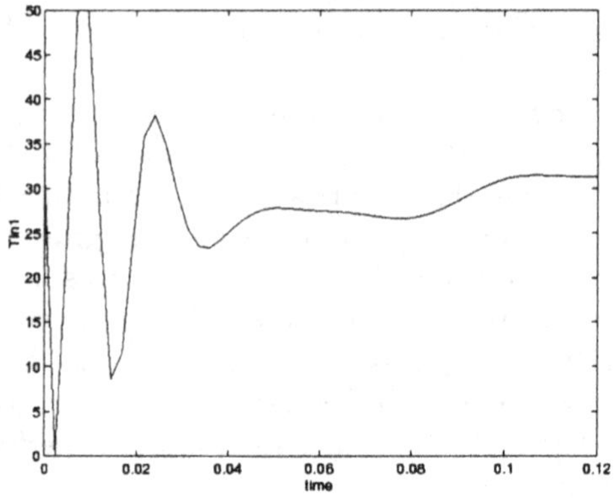

FIGURE 6.11. Example 5, optimal temperature—T_{in1}, short horizon.

5 Some Remarks on Integration and Optimization Accuracies

In all examples we applied the error estimates given by (3.11). Notice that error estimates were calculated only for differential states. At this point we would like to stress that formula (3.12) gives better error estimates but requires more computer memory (because *nd* is usually much smaller than *n*). Moreover, the integration time of systems equations is increased by 10%-20% but the number of steps can be decreased so that the integration time for adjoint equations can also decrease.

Notice that the norms in (3.5) and (3.13) (as well as that used to calculate ε in (3.5)) should be the same. However, we have observed that using the

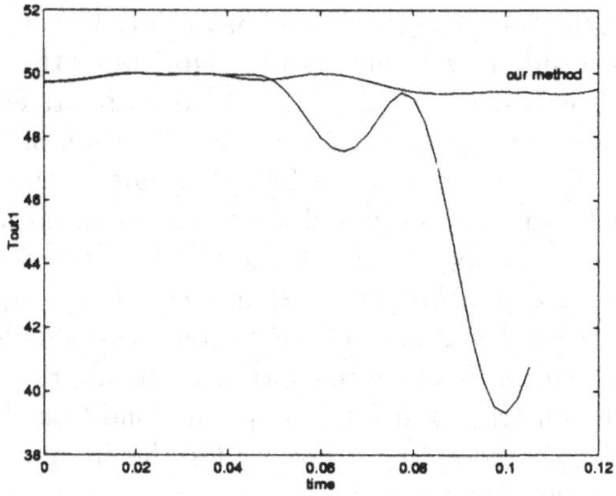

FIGURE 6.12. Example 5, optimal output—T_{out1}, comparison of sensitivity eqns. approach with our method.

l_2 norm in (3.13) and the l_∞ norm in (3.5) can reduce computational time.

In fact the norm used in (3.13) was a weighted root mean square norm (presented here for a vector $v \in \mathcal{R}^n$):

$$\|v\| = \sqrt{\frac{1}{n}\sum_{i=1}^{n}(v_i/WT_i)^2}$$

where

$$WT_i = \text{RTOL(I)}|v_i| + \text{ATOL(I)}$$

and RTOL(I) and ATOL(I) are relative and absolute tolerances for the ith component v_i. All calculations presented in this section were done with relative and absolute tolerances for differential states equal to 10^{-3} (with the exception of the first run in Example 5 in which case they were equal to 10^{-4}). It seems odd that the accuracy obtained with respect to constraints is much higher than the specified accuracy for the integration. This can be explained by the fact that continuous time differential–algebraic equations are transformed to implicit discrete time equations and if integration steps coincide with the discretization points for controls (using the notation from §6.2: $\mathcal{D}_s \subset \mathcal{D}_u$) then the formulas for the reduced gradients provide their very accurate evaluations. The accuracy of the reduced gradients is determined, in this case, by the accuracy with which equations (2.11) are satisfied. This accuracy was equal 10^{-7} for all runs.

If the condition $\mathcal{D}_s = \mathcal{D}_u$ does not hold approximately, $|\mathcal{D}_s| \gg |\mathcal{D}_u|$, then we will no longer guarantee the satisfaction of the constraints with higher accuracy than that specified for the integration. In this case we must be very careful when setting up tolerances for the constraints. They must reflect the number of significant digits which can be achieved by the integrator for the component of the state present in the constraint. The careless approach to tolerances can result in a failure of the algorithm—the approximation to the directional derivative, $\sigma^{DAE}_{c_l,A_l,H_l}(\vec{u}_l)$, assumes relatively big negative values but the corresponding direction is not a direction of descent for the penalty function. We recommend the following rule of thumb for setting up tolerances for the optimization and the integration: initially tolerances for constraints should be higher than those for the integration. If at some iteration, the optimization program requires small step α_l the program should be stopped and the accuracy for the integration should be increased and optimization continued. When the optimization program finds a control which satisfies all constraints and the stopping condition is met with the initially specified tolerances, the accuracy for the integration should be increased and the program run again.

Our main stopping criteria for the optimization algorithm is the satisfaction of all constraints with the specified accuracy and the condition that the approximation to the directional derivative of the exact penalty function, $\sigma^{DAE}_{c_k,A_k,H_k}(\vec{u}_k)$, is greater than some prespecified negative number. The latter condition reflects the fact that the value of the exact penalty function will not change much if we continue computations. It, however, does not measure how far the current controls are from a solution. Therefore, the stopping criterion based on the directional derivative should be supplemented by the additional requirement:

$$\begin{aligned} |(d_k(j))_i| &\leq \mathrm{RTOL_u(I)}|(u_k(j))_i| + \mathrm{ATOL_u(I)} \\ &\quad \forall i = 1, \ldots, m, \ \forall j = 0, \ldots, N-1, \end{aligned}$$

where $\mathrm{RTOL_u(I)}$ and $\mathrm{ATOL_u(I)}$ are relative and absolute tolerances for the ith component of a control vector. The criterion based on relative tolerances is especially recommended for problems described by large–scale differential–algebraic equations when fulfilling the requirement expressed only by absolute tolerances can result in unnecessary excessive computing time.

6 Concluding Remarks

The chapter presents a new numerical approach to optimal control of differential–algebraic equations. The numerical results suggest that it is more efficient than any other methods proposed for the problem. Moreover it can cope with problems described by large–scale systems of differential–algebraic equations. For example, if the last example (with $N = 100$) were attempted by the method based on the full discretization ([117]) we would end up with the optimization problem that would consist of more than half a million of decision variables (if the 2–stage Radau IIA integrator is applied and the discretization for controls is $N = 100$).

The implementation of the full discretization approach used in numerical comparisons presented in the previous section was based on an active set approach for large–scale nonlinear programming problems. It was an algorithm based on the generalized reduced gradient approach ([1]) which exploited a staircase structure of discrete time differential–algebraic equations ([38]). Using an interior–point method instead of an active set method should improve the efficiency of the full parametrization approach. Such a method was proposed by Wright ([126],[127]). Wright's approach assumes *a priori* discretization of system equations—thus the optimal control problem is transformed to a large–scale nonlinear programming problem. Then a direction finding subproblem is defined as a large–scale quadratic programming problem. This is essentially an accessory problem for discrete time optimal control problems based on second order derivatives. The accessory problem is solved by the interior–point method which fully exploits a staircase structure of the constraints. The way the interior–point method solves the accessory problem assumes that the Lagrange multipliers for the constraints are also independent variables. This means that the number of variables for the interior–point algorithm is equal to $N_s(4n + m + n_c)$ where n is the number of states, m the number of controls, n_c the number of constraints at each of N_s stages (if the 2–stage Radau IIA is used as an integrator). This implies that the cost of algebraic operations of one iteration of the interior–point method can be approximated (very pessimistically for large systems with sparse Jacobians) by $O(N_s(4n + m + n_c)^3)$ if the staircase structure of the constraints is taken into account in factorizations. ($O(k)$ means that there exists M such that $O(k) \leq Mk$.)

The cost of one iteration of the method described in this chapter can be very roughly approximated by $O((Nm)^2) + O(N_s(2n)^3) + O(N_s(2n)^2) + O(mNN_s(2n)^2)$.[7] (This estimate is valid if we assume that the number of iterations in $\mathbf{P}^{\mathbf{DAE}}_{\mathbf{c,A,H}}(\vec{\mathbf{u}})$ is moderate and the number of adjoint equations solved at each iteration is equal to mN which is a conservative assumption—see Table 6.6.) If $n > N_s > N$ then the approach proposed by Wright can be several times more expensive in terms of computing time.

---

[7]The estimate of the complexity of the LU factorization and the backward substitution is not very well worked out for systems with sparse Jacobians—[39]. However, in this case also we could show that the complexity of our algorithm is lower than that of the algorithm proposed by Wright (assuming that his factorization technique could be adapted to sparse matrices). This follows from the fact that the number of equations is two times higher in the Wright's approach and that the backward substitution is typically several times cheaper than the LU factorization—as a result the cost of the adjoint equations evaluation is comparable to the cost of the system equations integration.

Another drawback of the approach based on the interior–point method is the fact that the number of stages N_s must be predetermined. This can imply unnecessarily big N_s because we do not know *a priori* which number N_s guarantees the feasibility of implicit discrete time system equations. The approach introduced in this paper adjusts the number of stages N_s automatically to guarantee feasibility of the numerical integration. This procedure when combined with a good stepsize selection procedure usually results in relatively small numbers N_s.

Our method of handling state constraints can be combined with the evaluation of gradients based on sensitivity equations. Poor results recorded by the program proposed in [119] can be blamed, to some extent, by the fact that the program substitutes state constraints by the equality constraint (4.1). If this constraint is present in the formulation of an optimal control problem we cannot guarantee that an optimization algorithm will find a control satisfying necessary optimality conditions in a normal form, i.e.,the conditions stated in *Theorem 1* with $\alpha_0 = 1$. The sensitivity equations should be stated for systems equations (2.11) and their use is primarily recommended when the number of parameters to be optimized is relatively low.

Appendix A

A Primal Range–Space Method for Piecewise–Linear Quadratic Programming

A.1 Software Implementation

Software development, related to various numerical algorithms presented in the monograph, evolved during several years.

The first algorithm implemented was the feasible directions method discussed in Chapter 3 ([94]). In its realization we paid special attention to two crucial parts of an implementation of any optimal control algorithm: a quadratic programming procedure and an integration solver for the evaluation of system and adjoint equations.

The QP procedure employed in the first order method is described at length in the subsequent sections. Here, it is worthwhile to mention that we tried several approaches to the calculation of search directions. The first attempt was a proximity algorithm proposed for optimal control algorithms ([73], [123], [75]). However, we were very disappointed with its performance even though we implemented its efficient version with Hauser's improvements ([59]).[1] The performance of the proximity algorithm strongly depended on the number of general constraints in quadratic problems but even the moderate number of them (obviously problem dependent but, in most cases we tested, less than 30) resulted in the number of proximity algorithm iterations running in several hundred. We also tested the LSSOL code ([47]) but we soon realized that a null–space realization (see description in §A.2) of an active set method couldn't cope with large–scale search directions problems arising in the first order method. Eventually we implemented a range–space method—the subject of the appendix is to justify the choice.

The integration procedure employed in the PH2SOL program is the backward differentiation formula as realized in the LSODE code ([60]). The integration procedure was subsequently superseded by the implementation of Radau IIA Runge–Kutta procedure described in Chapter 6 with the aim of addressing large–scale differential–algebraic equations.

The Runge–Kutta procedure is at the heart of the RKCON program ([101]) for optimizing systems of differential–algebraic equations. The other

---

[1] Proximity algorithms can still be useful in solving convex problems with nondifferentiable objective functions as shown in [91] and [103].

part of the RKCON package is the second order algorithm introduced in Chapter 5 and coded as the SQPCON program ([99]). To facilitate handling large amount of data the RKCON has the interface to the gPROMS simulation package developed in the Centre for Process Systems Engineering at Imperial College ([8]).[2] The role of the interface is to adapt FORTRAN subroutines (generated by gPROMS) of system equations residuals and their Jacobians to the Runge–Kutta integrator.

The RKCON program is written in FORTRAN, in *double precision*, and uses BLAS subroutines whenever possible.

A.2 A Range–Space Method—Introduction

In the appendix we concentrate on an algorithm for a certain quadratic programming problem which arises when either a first, or second order method described in this monograph is applied to solve an optimal control problem. It is also an essential part of several algorithms of nondifferentiable optimization ([64]).

The problem of concern is the convex piecewise–linear quadratic programming problem with simple bounds on the variables—(**PLQP**):

$$\min_{x \in \mathcal{R}^n} \left[c^T x + 1/2 x^T H x + \max\left[0, \max_{i \in I} \left|a_i^T x + b_i\right|, \max_{j \in J}\left[a_j^T x + b_j\right]\right]\right] \tag{A.2.1}$$

$$\text{s. t.} \quad l \leq x \leq u, \tag{A.2.2}$$

where c, a_i, l, u are constant vectors, H is a constant $n \times n$ symmetric positive definite matrix and I, J are finite sets of indices.

The procedure proposed here can be applied to the problems (**PLQP**) with diagonal matrices with several thousand nonzero elements, if it is used in a first order method described in Chapters 3–4. In this case the method can be very efficient since it takes advantage of a diagonal structure of the Hessian matrix H. Alternatively it can be used in the SQP algorithm introduced in Chapter 5. In that case H is dense but with a moderate dimension nor exceeding, say, 500.

The procedure we propose for the problem (**PLQP**) is an active set method. Most active set algorithms for related QP problems can be classified as either *range-space* or *null-space* methods. This terminology arises because the set of constraints regarded as active (the working set) define two complementary subspaces: the *range space* of vectors that can be expressed as linear combinations of the associated vectors a_i, and the *null*

[2]The interface requires only minor changes to adapt it to the SPEEDUP simulation program ([114]).

space of vectors orthogonal to the vectors a_i. In many cases the work required in an iteration is directly proportional to the dimension of either the range space or the null space. Null–space methods are the most efficient when the number of constraints in the working set is close to n, contrary to range–space methods which are most efficient when there are few constraints.

The method described in the appendix is a range–space method. A feature of the method is that it is able to exploit the special structure of the objective function (A.2.1) and it copes efficiently with bound constraints (A.2.2). The method is an extension of the algorithm described in [46] and is largely based on the Gram–Schmidt orthogonalization presented in [35]. The method can also be applied to quadratic programming problems if an exact penalty function is applied to general linear constraints.

The method is a primal method which means that it generates a sequence of points $\{x_k\}$ feasible with respect to constraints (A.2.2). The problem (A.2.1)–(A.2.2) can also be solved by a dual algorithm ([65], [112]). Advantages (and disadvantages) of using a primal method over a dual one are discussed in §A.6.

A.3 The Basic Method

To simplify the presentation consider a simpler problem with the objective function

$$F(x) = c^T x + 1/2x^T Hx + \max_{i \in I} \left[a_i^T x + b_i\right] \tag{A.3.1}$$

and without simple constraints (A.2.2). Throughout, we assume that the matrix whose rows are vectors a_i, $i \in I$ has a full rank.

The problem (A.3.1) can be solved by introducing the additional variable β:

$$\min_{(x,\beta)} \left[c^T x + 1/2x^T Hx + \beta\right] \tag{A.3.2}$$

$$\text{s. t. } a_i^T x + b_i \leq \beta \;\forall i \in I. \tag{A.3.3}$$

The problem (A.3.2)–(A.3.3) is a QP problem with a positive semidefinite Hessian matrix and therefore the range–space algorithm described in [46] cannot be applied to it. Here we show that the algorithm from [46] can be modified to accommodate the singular Hessian of the problem (A.3.2)–(A.3.3).

The method we present here is an active set algorithm and that means that the solution to the problem is obtained by solving a sequence of quadratic programming problems with only equality constraints. If x_k is the current point generated by the algorithm then values of all piecewise–linear terms are evaluated

$$l_i(x_k) = a_i^T x_k + b_i$$

to define the set of active terms

$$I(x_k) = \left\{ i \in I : \max_{j \in I} l_j(x_k) = l_i(x_k) \right\}$$

and their maximum value

$$\max_{i \in I} l_i(x_k) = \beta_k.$$

The next point x_{k+1} is found under the assumption that the set of active terms does not change. To this end we solve the following quadratic programming problem in which the variable π corresponds to the update of the value of active terms. We call this problem $\mathbf{P}(x_k, J_k)$:

$$\min_{p,\pi} \left[c^T(x_k + p) + 1/2(x_k + p)^T H(x_k + p) + \beta_k + \pi \right] \tag{A.3.4}$$

$$\text{s. t.} \quad a_i^T(x_k + p) + b_i = \beta_k + \pi \quad \forall i \in J_k. \tag{A.3.5}$$

In order to solve problem (A.3.4)–(A.3.5) we have to find the solution to

$$\min_{p,\pi} \left[g_k^T p + 1/2 p^T H p + \pi \right] \tag{A.3.6}$$

$$\text{s. t.} \quad a_i^T p = \pi \quad \forall i \in J_k, \tag{A.3.7}$$

where $g_k = Hx_k + c$.

If (p_k, π_k) is the solution to problem (A.3.6)–(A.3.7) with $J_k = I(x_k)$ and $||p_k|| = 0$ (also $\pi_k = 0$ due to (A.3.7)) then either x_k is the solution to problem (A.3.2)–(A.3.3), or the set of active terms is different from that at the solution. In order to verify that the Lagrange multipliers λ_k of (A.3.6)–(A.3.7) are calculated. If all multipliers are nonnegative then x_k solves problem (A.3.2)–(A.3.3), otherwise removing a term j with the corresponding negative Lagrange multiplier guarantees that the optimal value of the problem $\mathbf{P}(x_k, J_k \setminus \{j\})$ is not greater than that for the problem $\mathbf{P}(x_k, J_k)$. Furthermore, if $(\bar{p}_k \neq 0, \bar{\pi}_k)$ is the solution to the problem $\mathbf{P}(x_k, J_k \setminus \{j\})$ then $a_j^T \bar{p}_k < \bar{\pi}_k$. Therefore, we set $J_k = J_k \setminus \{j\}$ in this case.

If $||p_k|| \neq 0$ we calculate the biggest possible step $0 < t_k^{max} \leq 1$ such that $J_k \subset I(x_k + t_k^{max} p_k)$. To this end we calculate

$$t_k^i : \quad a_i^T(x_k + t_k^i p_k) + b_i = \beta_k + t_k^i \pi_k \quad \forall i \notin I(x_k) \tag{A.3.8}$$

and we set t_k^{max} as the minimal value of all positive t_k^i:

$$t_k^{max} = \min \left[\{ t_k^i : i \notin I(x_k), \; t_k^i > 0 \} \bigcup \{1\} \right]. \tag{A.3.9}$$

The next point is given by $x_{k+1} = x_k + t_k^{max} p_k$ and at the point x_{k+1} we either solve the problem $\mathbf{P}(x_{k+1}, J_k \bigcup_j \{i_j\})$, where i_j are defined by the relation $t_k^{max} = t_k^{i_j}$, or the problem $\mathbf{P}(x_{k+1}, J_k)$ if $t_k^{max} = 1$ and $t_k^i \neq 1 \ \forall i \notin I(x_k)$.

We can prove that the sequence $\{x_k\}$ converges, in the finite number of iterations, to the solution of problem (A.3.2)–(A.3.3) (see, e.g, [14], [43], [48]). This follows from the fact that the algorithm performs iterations identical to those of a standard primal active set method applied to problem (A.3.2)–(A.3.3).

If vectors a_i, $i \in I(x_k)$ define the rows of the matrix A (where for the simplicity of the notation we drop a subscript k), then the solution pair (p, π) and the Lagrange multiplier λ of the problem $\mathbf{P}(x, I(x))$ are given as the solution to the equations

$$\begin{aligned} Hp + A^T\lambda + g &= 0 \quad &\text{(A.3.10)}\\ Ap &= \pi e \quad &\text{(A.3.11)}\\ e^T\lambda &= 1, \quad &\text{(A.3.12)} \end{aligned}$$

where $e = (1, 1, ..., 1) \in \mathcal{R}^m$, $m = |I(x)|$.

Knowing that A is of full rank we can solve the above equations analytically:

$$p = -H^{-1}\left[A^T\lambda + g\right] \quad \text{(A.3.13)}$$

$$\lambda = -\left[AH^{-1}A^T\right]^{-1}\left[\pi e + AH^{-1}g\right] \quad \text{(A.3.14)}$$

$$\pi = -\left(1 + e^T\left[AH^{-1}A^T\right]^{-1}AH^{-1}g\right)/e^T\left[AH^{-1}A^T\right]^{-1}e. \quad \text{(A.3.15)}$$

To solve equations (A.3.13)–(A.3.15) the following factorization of $R^{-T}A^T$, where R is the Cholesky factor of H ($R^TR = H$), is introduced (c.f. [7])

$$R^{-T}A^T = Q\begin{pmatrix} T \\ 0 \end{pmatrix}, \quad \text{(A.3.16)}$$

with T an $m \times m$ upper triangular matrix, and Q an $n \times n$ orthogonal matrix partitioned as

$$Q = (Y\ Z), \quad \text{(A.3.17)}$$

where Y, Z are $n \times m$ and $n \times (n - m)$ matrices.

Factorization (A.3.16) and equations (A.3.13)–(A.3.15) lead to the equations:

$$\begin{aligned} T\lambda &= -\ Y^TR^{-T}g - \pi T^{-T}e \quad &\text{(A.3.18)}\\ Rp &= YY^TR^{-T}g - R^{-T}g + \pi YT^{-T}e. \quad &\text{(A.3.19)} \end{aligned}$$

If we introduce the variables u, v, w, f, z in the following way

$$
\begin{aligned}
R^T u &= g && \text{(A.3.20)}\\
v &= Y^T u && \text{(A.3.21)}\\
w &= Y v - u && \text{(A.3.22)}\\
T^T f &= e && \text{(A.3.23)}\\
z &= Y f, && \text{(A.3.24)}
\end{aligned}
$$

then equation (A.3.18) is transformed to

$$T\lambda = -v - \pi f, \quad \text{(A.3.25)}$$

while equation (A.3.19) becomes

$$Rp = w + \pi z. \quad \text{(A.3.26)}$$

Moreover, using equation (A.3.25) and relation (A.3.12), we can finally provide the formula for π:

$$\pi = -\left(1 + f^T v\right) / \|f\|^2 . \quad \text{(A.3.27)}$$

We can now state the algorithm.

PLQP Algorithm

Step 0. Initialization.

0.1. Choose an initial vector: Set x, $J = I(x)$ and calculate g.

0.2. Initiate matrices and vectors: Initiate R, Y, T, vectors u, v, w, f, z and scalar π.

Step 1. Solve the problem $\mathbf{P}(x, J)$.

1.1. Find a new direction of descent: Solve equations (A.3.26) for p.

1.2. Check optimality of the problem $\mathbf{P}(x, J)$*:* If $\|p\| \neq 0$ then p is not optimal for $\mathbf{P}(x, J)$ and go to *Step 4*.

Step 2. Optimality test.

2.1. Find Lagrange multipliers: Solve equations (A.3.25) for λ.

2.2. Check optimality: If $\lambda \geq 0$ then x solves problem (A.3.1) and STOP.

Step 3. Active set reduction.

3.1. Choose a constraint to drop: Find $j \in J$ with the negative Lagrange multiplier $\lambda^j < 0$. Modify the active set by removing j from J.

3.2. Modify matrices and vectors: Modify matrices Y, T, vectors u, v, w, f, z and scalar π. Go to *Step 1.*

Step 4. Update active set.

4.1. Calculate the stepsize: Find the largest possible step $t^{max} > 0$ according to (A.3.8)–(A.3.9).

4.2. Add constraint to active set: If there exist $j \notin I(x)$ such that $t^{max} = t^j$, augment J by these j.

4.3 Modify matrices and vectors: Substitute $x + t^{max}p$ for x, $g + t^{max}Hp$ for g, modify matrices Y, T, vectors u, v, w, f, z and scalar π. Go to *Step 1.*

A.4 Efficient Implementation

The basic algorithm presented in the previous section should be modified in order to accommodate simple constraints (A.2.2). We follow [46] to accomplish it. However, several modifications are required in the formulas for the updates of R, Y, T, u, v, f and z.

The simple constraints are a particular case of the general constraints

$$(a_j^0)^T x - b_j^0 \leq 0 \quad \forall j \in I^0. \tag{A.4.1}$$

The problem with the objective function (A.3.1) and constraints (A.4.1) has a direction finding subproblem defined by equations similar to (A.3.10)–(A.3.12). If we denote by $I^0(x)$ the indices of constraints (A.4.1) active at x, by A^0 the matrix whose rows are vectors a_j^0, $j \in I^0(x)$ then equations (A.3.10)- (A.3.12) have to be replaced by

$$Hp + A^T\lambda + (A^0)^T\lambda^0 + g = 0 \tag{A.4.2}$$
$$A^0 p = 0 \tag{A.4.3}$$
$$Ap = \pi e \tag{A.4.4}$$
$$e^T\lambda = 1, \tag{A.4.5}$$

where, as before, $e = (1, 1, ..., 1) \in \mathcal{R}^m$, $m = |I(x)|$. The relations (A.3.20)–(A.3.24) have to be modified accordingly, for example vector e in (A.3.23) will have some elements, those corresponding to the rows of the matrix A^0, equal to zero.

At an iteration of the active set method applied to the PLQP problem, the working set will include a mixture of vectors a_i, $i \in I(x)$ and bounds. If the working set includes simple bounds, these variables will be *fixed* on the corresponding bounds during the given iteration; all other variables are considered as free to *vary*. We use the suffixes 'F' ('fixed') and 'V' ('varying') to denote items associated with the two types of variables.

If constraints (A.4.1) are simple constraints (A.2.2) then there is no need to use equations (A.4.2)–(A.4.5) but equations (A.3.10)–(A.3.12) for free variables only. As it was noted in [46], it is more convenient to use the weighted Gram–Schmidt factorization instead of (A.3.16):

$$R^{-T}A^T = Y\hat{T}, \tag{A.4.6}$$

where $\hat{T}$ is nonsingular, but not necessarily upper triangular matrix, and Y is a basis for the row space $R^{-T}A^T$. Relations (A.3.20)–(A.3.27) are still valid for this factorization of $R^{-T}A^T$. The weighted Gram–Schmidt factorization (A.4.6) is used in the rest of this section.

We denote by C the $t \times n$ matrix whose rows are defined by bound constraints and vectors a_i, $i \in I(x)$. We assume that C contains n_F bounds and m_L vectors a_i, $i \in I(x)$ ($t = n_F + m_L$, $m_L = |I(x)|$), and that all variables are ordered so that the free variables come first. Thus,

$$C = \begin{pmatrix} 0 & I_F \\ A_V & A_F \end{pmatrix},$$

where the matrix A is divided accordingly $A = (A_V \; A_F)$ and, for the simplicity of presentation, I_F is the identity matrix with rank n_F (in general I_F is a diagonal matrix with diagonal elements equal to 1 or -1). We assume that C has full rank.

If the box constraints are taken into account in direction finding subproblem (A.4.2)–(A.4.5), equations (A.4.3)–(A.4.4) should be substituted by the following equations

$$I_F p_F = 0 \tag{A.4.7}$$

$$A_V p_V + A_F p_F = \pi e, \tag{A.4.8}$$

where $p = (p_V, p_F)$ is divided according to the definition of 'varying' and 'fixed' variables.

The Cholesky factor R is also rearranged in a similar way

$$R = \begin{pmatrix} R_V & S \\ 0 & R_F \end{pmatrix},$$

where R_V and R_F are $n_V \times n_V$ and $n_F \times n_F$ upper triangular matrices.

If the Gram–Schmidt factorization of $R_V^{-T}A_V^T$ is known,

$$A_V^T = R_V^T Y_V T_L,$$

where T_L is an $m_L \times m_L$ upper-triangular matrix, Y_V is an $n_V \times m_L$ orthonormal matrix, then the weighted Gram–Schmidt factorization of the matrix $R^{-T}C^T$ is given by

$$C^T = R^T \begin{pmatrix} 0 & Y_V \\ I_F & 0 \end{pmatrix} \begin{pmatrix} N & M \\ 0 & T_L \end{pmatrix} = R^T Y T,$$

where N, M are matrices of order $n_F \times n_F$ and $n_F \times m_L$ respectively. Note that T is not an upper triangular matrix and that the matrix

$$Y = \begin{pmatrix} 0 & Y_V \\ I_F & 0 \end{pmatrix}$$

forms a basis for the row space $R^{-T}C^T$. Observe also that it is not necessary to store the matrices N, M because the matrices A_V, R_V, T_L, Y_V contain the information necessary to perform the iteration. We can use these matrices and relations (A.3.20)–(A.3.27) to calculate a new point x.

If g and u are partitioned as

$$g = \begin{pmatrix} g_V \\ g_F \end{pmatrix}, \quad u = \begin{pmatrix} u_V \\ u_F \end{pmatrix}$$

then, from the definition of R, $g_V = R_V^T u_V$ and

$$v = Y^T u = \begin{pmatrix} 0 & I_F \\ Y_V^T & 0 \end{pmatrix} \begin{pmatrix} u_V \\ u_F \end{pmatrix} = \begin{pmatrix} u_F \\ Y_V^T u_V \end{pmatrix} = \begin{pmatrix} u_F \\ v_L \end{pmatrix},$$

where $v_L \in \mathcal{R}^{m_L}$.

In a similar way we can show that

$$w = Yv - u = \begin{pmatrix} w_V \\ 0 \end{pmatrix}.$$

Moreover

$$\begin{pmatrix} N^T & 0 \\ M^T & T_L^T \end{pmatrix} \begin{pmatrix} f_F \\ f_L \end{pmatrix} = \begin{pmatrix} 0 \\ e \end{pmatrix},$$

where $0 \in \mathcal{R}^{n_F}$ and $e \in \mathcal{R}^{m_L}$. From this equation, because N is nonsingular due to our assumption that C has full rank, we have $f_F = 0$, therefore

$$z = \begin{pmatrix} 0 & Y_V \\ I_F & 0 \end{pmatrix} \begin{pmatrix} 0 \\ f_L \end{pmatrix} = \begin{pmatrix} Y_V f_L \\ 0 \end{pmatrix} = \begin{pmatrix} z_V \\ 0 \end{pmatrix}.$$

This means that $p = (p_V^T \; 0)^T$ where p_V is the solution of the equation

$$R_V p_V = w_V + \pi z_V. \tag{A.4.9}$$

Furthermore, π is calculated according to the formula

$$\pi = -(1 + f_L^T v_L)/\left\|f_L\right\|^2.$$

In order to perform Step 2 of the *PLQP Algorithm*, adapted for the problem (**PLQP**), we also need to calculate the Lagrange multipliers corresponding to the problem $\mathbf{P}(x, J)$. There are two types of Lagrange multipliers. The first ones are associated with the bound constraints. Because

$p_F = 0$ (from (A.4.7)) we can write the following equation for these multipliers:

$$g_F + A_F^T \lambda_L + \lambda_F = 0. \tag{A.4.10}$$

We have the following equation for the Lagrange multipliers associated with the linear nondifferentiable term in the objective function:

$$R_V^T R_V p_V + A_V^T \lambda_L + g_V = 0.$$

If we premultiply this equation by $Y_V^T R_V^{-T}$ and take into account that $Y_V^T w_V = 0$, $R_V p_V = w_V + \pi z_V$, $Y_V^T z_V = f_L$, we will get

$$T_L \lambda_L + v_L + \pi f_L = 0.$$

Therefore, all crucial equations can be expressed by matrices: Y_V, R_V, A and vectors u_V, w_V, v_L, f_L, z_V, g_F.

In the rest of this section we will describe how R_V, Y_V, T_L, u_V, w_V, etc. are updated when a bound or a linear term is added or deleted from the working set. These updates are similar (but not identical) to those presented in [46]. These updates correspond to the following change of x: $\bar{x} = x + \alpha p$.

A.4.1 Adding a Bound to the Working Set

The procedure of adding a bound to the working set consists of two major steps. In the first step the jth variable $j \leq n_V$ is moved from its position to the last n_Vth position. This requires reordering of free variables. Let Π be the permutation matrix which represents this reordering. This matrix applied to the Cholesky factor R_V gives the upper Hessenberg matrix $R_V \Pi$. We can construct a sequence of the plane rotations $\{Q_{k,k+1}\}$ (where $Q_{k,k+1}$ changes k and $k+1$ rows of a given matrix) that transform $R_V \Pi$ to the upper triangular matrix:

$$\hat{R}_V = Q_{n_V-1,n_V} \cdots Q_{j+1,j+2} Q_{j,j+1} R_V \Pi = Q R_V \Pi. \tag{A.4.11}$$

The matrix $\hat{R}_V$ is the Cholesky factor of the reordered Hessian $\Pi^T H \Pi$

$$\hat{R}_V^T \hat{R}_V = \Pi^T R_V^T Q^T Q R_V \Pi = \Pi^T R_V^T R_V \Pi = \Pi^T H \Pi,$$

since Q is orthonormal.

These changes affect the orthonormal matrix Y_V according to the relation

$$\Pi^T A_V^T = \hat{A}_V^T = \Pi^T R_V^T Y_V T_L = \Pi^T R_V^T Q^T Q Y_V T_L = \hat{R}_V^T \hat{Y}_V T_L.$$

In the next step the last n_Vth variable becomes fixed. This corresponds to deleting the last column, say a, of the matrix $\hat{A}_V$. In this case $\bar{R}_V$ is simply a submatrix of $\hat{R}_V$

$$\begin{pmatrix} \bar{R}_V & r \\ 0 & \rho \end{pmatrix} = \hat{R}_V.$$

The Gram–Schmidt factorization of $\bar{R}_V^{-T}\bar{A}_V^T$ is computed as follows. Define $\tilde{Y}_V$ by the equation

$$\hat{R}_V^{-T}\hat{A}_V^T = \hat{Y}_V T_L = \begin{pmatrix} \tilde{Y}_V \\ y^T \end{pmatrix} T_L$$

which can also be stated as

$$\hat{R}_V^{-T}\hat{A}_V^T = \begin{pmatrix} \tilde{Y}_V & 0 \\ y^T & 1 \end{pmatrix}\begin{pmatrix} T_L \\ 0 \end{pmatrix}.$$

The Gram–Schmidt orthogonalization (with reorthogonalization if necessary —[35], p. 779) is then applied to give

$$\begin{pmatrix} \tilde{Y}_V & 0 \\ y^T & 1 \end{pmatrix} = \begin{pmatrix} \tilde{Y}_V & \tilde{y} \\ y^T & \sigma \end{pmatrix}\begin{pmatrix} I & t \\ 0 & \tau \end{pmatrix},$$

where vectors $\tilde{y}$, t and the scalar τ are defined by the Gram–Schmidt orthogonalization process ([35], p. 773). We can show that $t = y$, $\sigma = \tau$ ([35], p. 779), therefore

$$\begin{pmatrix} \tilde{Y}_V & 0 \\ y^T & 1 \end{pmatrix} = \begin{pmatrix} \tilde{Y}_V & \tilde{y} \\ y^T & \tau \end{pmatrix}\begin{pmatrix} I & y \\ 0 & \tau \end{pmatrix}.$$

Hence, we have

$$\hat{R}_V^{-T}\hat{A}_V^T = \begin{pmatrix} \tilde{Y}_V & \tilde{y} \\ y^T & \tau \end{pmatrix}\begin{pmatrix} T_L \\ 0 \end{pmatrix}.$$

We now choose Givens matrices P_{m_L,m_L+1}, $P_{m_L-1,m_L+1}, \ldots, P_{1,m_L+1}$ such that

$$(y^T\ \tau)P^T = (y^T\ \tau)P^T_{m_L,m_L+1}\cdots P^T_{1,m_L+1} = (0\ \omega).$$

Moreover, $\omega = \pm 1$ since P is an orthonormal transformation and preserves the length $(||y||^2 + \tau^2 = 1$—[35], p. 779). Therefore, we have

$$\begin{pmatrix} \tilde{Y}_V & \tilde{y} \\ y^T & \tau \end{pmatrix} P^T = \begin{pmatrix} \bar{Y}_V & \bar{y} \\ 0 & \omega \end{pmatrix} \tag{A.4.12}$$

and since this matrix has orthonormal columns: $\bar{y} = 0$.

The transformation P applied to the matrix T_L, enlarged by the zero vector, gives

$$P\begin{pmatrix} T_L \\ 0 \end{pmatrix} = P_{1,m_L+1}\cdots P_{m_L,m_L+1}\begin{pmatrix} T_L \\ 0 \end{pmatrix} = \begin{pmatrix} \bar{T}_L \\ \bar{t}^T \end{pmatrix},$$

where $\bar{T}_L$ is upper triangular.

Finally, we have

$$\hat{A}_V^T = \begin{pmatrix} \bar{A}_V^T \\ a^T \end{pmatrix} = \begin{pmatrix} \bar{R}_V^T & 0 \\ r^T & \rho \end{pmatrix} \begin{pmatrix} \bar{Y}_V & 0 \\ 0 & \omega \end{pmatrix} \begin{pmatrix} \bar{T}_L \\ \bar{t}^T \end{pmatrix}$$

and $\bar{A}_V^T = \bar{R}_V^T \bar{Y}_V \bar{T}_L$, as required.

The following theorem shows how to update vectors u_V, v_L, w_V, f_L and z_V when a bound is added to the working set (assume $\alpha = t^{max}$ in this case):

Theorem A.4.1 *Suppose the new variable is fixed on a bound at the point* $\bar{x} = x + \alpha p$. *Assume that the updated factors* $\bar{R}_V$, $\bar{T}_L$, $\bar{Y}_V$ *have been computed, then the vectors* u_V, v_L, w_V, f_L, z_V *are updated as follows:*

(i)

$$\begin{pmatrix} \bar{u}_V \\ \eta \end{pmatrix} = Q(u_V + \alpha[w_V + \pi z_V]); \qquad \text{(A.4.13)}$$

(ii)

$$\begin{pmatrix} \bar{v}_L \\ \omega\eta \end{pmatrix} = P \begin{pmatrix} v_L + \alpha\pi f_L \\ \nu \end{pmatrix}, \quad \nu = (1-\alpha)(\tilde{y}^T \ \tau) Q u_V; \qquad \text{(A.4.14)}$$

(iii)

$$\begin{pmatrix} \bar{w}_V \\ 0 \end{pmatrix} = Q(1-\alpha) w_V + \nu \begin{pmatrix} \tilde{y} \\ \tau \end{pmatrix}; \qquad \text{(A.4.15)}$$

(iv)

$$\begin{pmatrix} \bar{f}_L \\ 0 \end{pmatrix} = P \begin{pmatrix} f_L \\ \mu \end{pmatrix},$$

where μ *is the scalar which can be calculated from* f_L *and components of the matrix* P;

(v)

$$\begin{pmatrix} \bar{z}_V \\ 0 \end{pmatrix} = Q z_V + \mu \begin{pmatrix} \tilde{y} \\ \tau \end{pmatrix}.$$

Proof.

In order to prove *(i)* first of all we apply the permutation to the gradient at the point $\bar{x} = x + \alpha p$:

$$\tilde{g}_V = \begin{pmatrix} \bar{g}_V \\ \gamma \end{pmatrix} = \Pi^T (g_V + \alpha R_V^T R_V p_V).$$

If we take into account the relations: $R_V^T u_V = g_V$, $R_V p_V = w_V + \pi z_V$ we will get

$$\begin{pmatrix} \bar{g}_V \\ \gamma \end{pmatrix} = \Pi^T R_V^T (u_V + \alpha[w_V + \pi z_V]). \tag{A.4.16}$$

Since $\bar{R}_V^T \bar{u}_V = \bar{g}_V$ we can write

$$\begin{pmatrix} \bar{R}_V^T & 0 \\ r^T & \rho \end{pmatrix} \begin{pmatrix} \bar{u}_V \\ \eta \end{pmatrix} = \begin{pmatrix} \bar{g}_V \\ \gamma \end{pmatrix}.$$

Using (A.4.11) and (A.4.16) we come to

$$\Pi^T R_V^T Q^T \begin{pmatrix} \bar{u}_V \\ \eta \end{pmatrix} = \Pi^T R_V^T (u_V + \alpha[w_V + \pi z_V])$$

and therefore

$$\begin{pmatrix} \bar{u}_V \\ \eta \end{pmatrix} = Q(u_V + \alpha[w_V + \pi z_V]), \tag{A.4.17}$$

since Q is orthonormal. This proves *(i)*.

In order to prove *(ii)* we note that

$$\begin{pmatrix} \bar{v}_L \\ \omega\eta \end{pmatrix} = \begin{pmatrix} \bar{Y}_V^T & 0 \\ 0 & \omega \end{pmatrix} \begin{pmatrix} \bar{u}_V \\ \eta \end{pmatrix}.$$

If we take into account (A.4.12) and (A.4.17), we will get

$$\begin{aligned} \begin{pmatrix} \bar{v}_L \\ \omega\eta \end{pmatrix} &= P \begin{pmatrix} \tilde{Y}_V^T & y^T \\ \tilde{y}^T & \tau \end{pmatrix} Q(u_V + \alpha[w_V + \pi z_V]) \\ &= P \begin{pmatrix} Y_V^T \\ (\tilde{y}^T \ \tau)Q \end{pmatrix} (u_V + \alpha[w_V + \pi z_V]). \end{aligned}$$

Since $Y_V^T w_V = 0$, $Y_V^T z_V = f_L$, $Y_V^T u_V = v_L$ and $(\tilde{y}^T \ \tau)^T$ is orthonormal to QY_V we have proved *(ii)*.

From the definition of $\bar{w}_V$ we have:

$$\begin{pmatrix} \bar{w}_V \\ 0 \end{pmatrix} = \begin{pmatrix} \bar{Y}_V & 0 \\ 0 & \omega \end{pmatrix} \begin{pmatrix} \bar{v}_L \\ \omega\eta \end{pmatrix} - \begin{pmatrix} \bar{u}_V \\ \eta \end{pmatrix} =$$

$$\left(QY_V \ \begin{pmatrix} \tilde{y} \\ \tau \end{pmatrix} \right) \begin{pmatrix} v_L + \alpha\pi f_L \\ \nu \end{pmatrix} - Q(u_V + \alpha[w_V + \pi z_V]) =$$

$$Q(Y_V v_L + \alpha\pi Y_V f_L - u_V - \alpha w_V - \alpha\pi z_V) + \nu \begin{pmatrix} \tilde{y} \\ \tau \end{pmatrix}.$$

which gives the desired result *(iii)*.

$\bar{f}_L$ has to satisfy the equation

$$\begin{aligned}(\bar{T}_L^T \; t)\begin{pmatrix}\bar{f}_L\\0\end{pmatrix} &= e\\ &= (T_L^T \; 0)P^T\begin{pmatrix}\bar{f}_L\\0\end{pmatrix}.\end{aligned} \tag{A.4.18}$$

Denote $(s^T, \mu)^T$ as follows

$$P^T\begin{pmatrix}\bar{f}_L\\0\end{pmatrix} = \begin{pmatrix}s\\\mu\end{pmatrix}.$$

Then, from equation (A.4.18), s must be equal to f_L because T_L is non-singular. Therefore, in order to calculate $\bar{f}_L$ we have to find μ. Since P is orthonormal, $\bar{f}_L$ depends on μ according to the relation

$$P\begin{pmatrix}f_L\\\mu\end{pmatrix} = \begin{pmatrix}\bar{f}_L(\mu)\\t(\mu)\end{pmatrix}.$$

If $t(\mu)$ depends on μ then μ can be calculated from the equation $t(\mu) = 0$. Otherwise we can show that at least one element of $\bar{f}_L(\mu)$ is dependent on μ and therefore μ can be calculated from one of the equations in

$$\bar{T}^T \bar{f}_L(\mu) = e.$$

Alternatively, in the less efficient implementation, μ can be found by solving equations (A.4.18) for $\bar{f}_L$. Then μ is calculated from the relation $\bar{f}_L = \bar{f}_L(\mu)$, or $\bar{z}_V$ is updated directly from the relation $\bar{z}_V = \bar{Y}_V \bar{f}_L$. This completes the proof of part *(iv)*.

From the definition of $\bar{z}_V$ and (A.4.12) we obtain:

$$\begin{aligned}\begin{pmatrix}\bar{z}_V\\0\end{pmatrix} &= \begin{pmatrix}\bar{Y}_V & 0\\0 & \omega\end{pmatrix}\begin{pmatrix}\bar{f}_L\\0\end{pmatrix}\\ &= \left(QY_V \; \begin{pmatrix}\tilde{y}\\\tau\end{pmatrix}\right)P^T P\begin{pmatrix}f_L\\\mu\end{pmatrix}\\ &= Qz_V + \mu\begin{pmatrix}\tilde{y}\\\tau\end{pmatrix},\end{aligned}$$

therefore the part *(v)* is proved. ■

A.4.2 Deleting a Bound from the Working Set

When the (n_V+j)th bound is deleted from the working set, the new variable is placed in the last (n_V+1)th position. Therefore R becomes

$$\tilde{R} = \begin{pmatrix}R_V & \tilde{r}_1 & \tilde{S}\\0 & \tilde{r}_2 & \tilde{R}_F\end{pmatrix}.$$

Plane rotations must be applied to the matrix

$$\tilde{R}_V = \begin{pmatrix} R_V & \tilde{r}_1 \\ 0 & \tilde{r}_2 \end{pmatrix}$$

so that it becomes upper triangular:

$$Q^1_{n_V+2,n_V+1} \cdots Q^1_{n_V+j,n_V+j-1} \begin{pmatrix} R_V & \tilde{r}_1 & \tilde{S} \\ 0 & \tilde{r}_2 & \tilde{R}_F \end{pmatrix} = \begin{pmatrix} R_V & r & \bar{S}_1 \\ 0 & \rho & \bar{s}^T \\ 0 & 0 & \hat{R}_F \end{pmatrix}. \tag{A.4.19}$$

Since $\bar{R}_F$ should be upper triangular, additional plane rotations are applied to the lower part of the matrix R. Eventually we obtain

$$Q^2_{n_V+j+1,n_V+j} \cdots Q^2_{n_V+3,n_V+2} \begin{pmatrix} R_V & r & \bar{S}_1 \\ 0 & \rho & \bar{s}^T \\ 0 & 0 & \hat{R}_F \end{pmatrix} = \begin{pmatrix} \bar{R}_V & \bar{S} \\ 0 & \bar{R}_F \end{pmatrix}. \tag{A.4.20}$$

Deleting a bound from the working set means that a column a must be added to the matrix A_V. The augmented matrix $\bar{A}_V$ may be written as

$$\bar{A}_V^T = \begin{pmatrix} A_V^T \\ a^T \end{pmatrix} = \bar{R}_V^T \begin{pmatrix} Y_V & 0 \\ 0 & 1 \end{pmatrix} \begin{pmatrix} T_L \\ t^T \end{pmatrix},$$

where $t = \bar{A}_V q$. q is a solution of the equation $\bar{R}_V q = i_{n_V+1}$ where i_{n_V+1} is the $(n_V + 1)$th coordinate vector.

$(T_L^T\ t)^T$ is upper triangular except for the 'horizontal spike' t. This spike can be reduced to the zero vector by the product P of the plane rotations $P_{1,m_L+1}, P_{2,m_L+1}, \ldots, P_{m_L,m_L+1}$:

$$\begin{aligned} \bar{A}_V^T &= \bar{R}_V^T \begin{pmatrix} Y_V & 0 \\ 0 & 1 \end{pmatrix} P^T P \begin{pmatrix} T_L \\ t^T \end{pmatrix} \\ &= \bar{R}_V^T (\bar{Y}_V\ \bar{y}) \begin{pmatrix} \bar{T}_L \\ 0 \end{pmatrix} \\ &= \bar{R}_V^T \bar{Y}_V \bar{T}_L \end{aligned} \tag{A.4.21}$$

as required.

Now we can state formulas for the updates of vectors u_V, v_L, w_V, f_L and z_V if a bound is deleted from the working set (assume $\alpha = 0$ in this case).

Theorem A.4.2 *Suppose that a bound is to be deleted from the working set at the point $\bar{x} = x + \alpha p$. Assume that the updated factors $\bar{R}_V$, $\bar{T}_L$, $\bar{Y}_V$ have been computed, then the vectors u_V, v_L, w_V, f_L, z_V are updated as follows:*

(i)

$$\bar{u}_V = \begin{pmatrix} u_V + \alpha[w_V + \pi z_V] \\ \mu \end{pmatrix}, \tag{A.4.22}$$

where μ can be calculated from u_V, g_F and some elements of $\bar{R}_V$;

(ii)

$$\begin{pmatrix} \bar{v}_L \\ \eta \end{pmatrix} = P \begin{pmatrix} v_L + \alpha\pi f_L \\ \mu \end{pmatrix};$$

(iii)

$$\bar{w}_V = \begin{pmatrix} (1-\alpha)w_V \\ 0 \end{pmatrix} - \eta\bar{y};$$

(iv)

$$\begin{pmatrix} \bar{f}_L \\ \nu \end{pmatrix} = P \begin{pmatrix} f_L \\ 0 \end{pmatrix};$$

(v)

$$\bar{z}_V = \begin{pmatrix} z_V \\ 0 \end{pmatrix} - \nu\bar{y}.$$

Proof.

The proof of the theorem is given in [95]. The proof, as the proofs of the subsequent theorem presented in this section, uses arguments highlighted in the proof of *Theorem A.4.1.* ■

A.4.3 Adding a Vector a to the Working Set

If a is added to the working set, Y_V and T_L must be changed to reflect a new vector $R_V^{-T}a = q$ in the matrix $R_V^{-T}\bar{A}_V^T$:

$$R_V^{-T}\bar{A}_V^T = (R_V^{-T}A_V^T \; R_V^{-T}a) = (Y_V \; q)\begin{pmatrix} T_L & 0 \\ 0 & 1 \end{pmatrix}.$$

We apply the Gram–Schmidt orthogonalization (with reorthogonalization if necessary) to the matrix $(Y_V\ q)$:

$$(Y_V\ q) = (Y_V\ y)\begin{pmatrix} I & t \\ 0 & \tau \end{pmatrix}.$$

Finally, we have

$$R_V^{-T}\bar{A}_V^T = (Y_V\ y)\begin{pmatrix} T_L & t \\ 0 & \tau \end{pmatrix} = \bar{Y}_V\bar{T}_L. \qquad \text{(A.4.23)}$$

The following theorem describes how the quantities u_V, v_L, w_V, f_L and z_V may be updated following the addition of a vector a (assume $\alpha = t^{max}$ in this case).

Theorem A.4.3 *Let $\bar{x} = x + \alpha p$ be a point at which the vector a is added to A_V. Assume that the updated factors $\bar{Y}_V$, $\bar{T}_L$ have been computed. The vectors u_V, v_L, w_V, f_L, z_V are updated as follows:*

(i)

$$\bar{u}_V = u_V + \alpha[w_V + \pi z_V]; \qquad \text{(A.4.24)}$$

(ii)

$$\bar{v}_L = \begin{pmatrix} v_L + \alpha\pi f_L \\ \nu \end{pmatrix}, \quad \nu = (1-\alpha)y^T u_V; \qquad \text{(A.4.25)}$$

(iii)

$$\bar{w}_V = (1-\alpha)w_V + \nu y; \qquad \text{(A.4.26)}$$

(iv)

$$\bar{f}_L = \begin{pmatrix} f_L \\ \mu \end{pmatrix}, \quad \mu = (1 - f_L^T t)/\tau;$$

(v)

$$\bar{z}_V = z_V + \mu y.$$

Proof.

The proof is given in [95]. ■

If more than one general constraint (or a bound) is added to the working set, formulas (A.4.13)–(A.4.15) and (A.4.24)–(A.4.26) for the updates remain valid with $\alpha = 0$.

A.4.4 Deleting a Vector a from the Working Set

The deletion of a from the working set means that some columns should be removed from Y_V and T_L:

$$(R_V^{-T} A_{1V}^T \;\; R_V^{-T} a \;\; R_V^{-T} A_{2V}^T) = Y_V\,(T_{1L} \;\; t \;\; T_{2L}).$$

Then

$$R_V^{-T}\,(A_{1V}^T \;\; A_{2V}^T) = Y_V (T_{1L} \;\; T_{2L}) = Y_V \hat{T}_L,$$

where $\hat{T}_L$ is the upper Hessenberg matrix. To transform $\hat{T}_L$ to an upper triangular form we use Givens rotations. We choose Givens matrices $G_{k,k+1}, G_{k+1,k+2}, \ldots, G_{m_L-1,m_L}$, where k is the position of a in the matrix A_V such that

$$G\hat{T}_L = G_{m_L-1,m_L} \cdots G_{k,k+1}\hat{T}_L = \begin{pmatrix} \bar{T}_L \\ 0 \end{pmatrix}$$

with $\bar{T}_L$ upper triangular. Furthermore,

$$Y_V G^T = Y_V G_{k,k+1} \cdots G_{m_L-1,m_L} = (\bar{Y}_V \;\; \bar{y}) \tag{A.4.27}$$

has orthonormal columns and $R_V^{-T}(A_{1V}^T \; A_{2V}^T) = \bar{Y}_V \bar{T}_L$ as required.

The following theorem shows how the quantities u_V, v_L, w_V, f_L and z_V change when a vector a is deleted from the working set (assume $\alpha = 0$ in this case).

Theorem A.4.4 *Suppose that the vector a is deleted from the working set at the point $\bar{x} = x + \alpha p$ and that the factors $\bar{Y}_V$, $\bar{T}_L$ have been computed. Then the vectors u_V, v_L, w_V, f_L, z_V are updated as follows:*

(i)

$$\bar{u}_V = u_V + \alpha[w_V + \pi z_V]; \tag{A.4.28}$$

(ii)

$$\begin{pmatrix} \bar{v}_L \\ \nu \end{pmatrix} = G[v_L + \alpha\pi f_L];$$

(iii)

$$\bar{w}_V = (1-\alpha)w_V - \nu\bar{y};$$

(iv)

$$\begin{pmatrix} \bar{f}_L \\ \mu \end{pmatrix} = G f_L; \tag{A.4.29}$$

(v)

$$\bar{z}_V = z_V - \mu\bar{y}.$$

Proof.

The proof is given in [95]. ∎

Eventually, when the *PLQP Algorithm*, adapted for the problem (**PLQP**), takes the Newton step ($t^{max} = 1$) without hitting any constraints then u_V, w_V, f_L, z_V and π change in the following way (we provide these formulas without proof since deriving them is straightforward)

$$\begin{aligned} \bar{u}_V &= u_V + w_V + \pi z_V \\ \bar{v}_L &= v_L + \pi f_L \\ \bar{w}_V &= 0 \\ \bar{f}_L &= f_L \\ \bar{\pi} &= 0. \end{aligned}$$

The number of multiplications required at an iteration of the *PLQP Algorithm* depends whether a bound or a linear term enters or is deleted from the working set. These numbers are very close to the estimates for a range–space method presented in [46] (p. 194).

The *PLQP Algorithm* significantly simplifies when the Hessian matrix H is diagonal. The equation (A.4.9) is then solved with the number of multiplications equal to n_V. Furthermore, the transformations to triangular matrices, in (A.4.11), (A.4.19) and (A.4.20), do not require plane rotations. This implies that in the case of the diagonal matrix H the number of multiplications is proportional to $n_V \times m_L + m_L^2$.

A.5 Computation of the Lagrange Multipliers Corresponding to the Fixed Variables

In order to delete a bound from the working set we have to know a part of gradient corresponding to the fixed variables: g_F. Having it we can state a formula for the Lagrange multipliers associated with the fixed variables (see (A.4.10)).

The gradient g_F will change in two situations: when we move from the point x to the point $\bar{x} = x + \alpha p$; when we change the status of a bound.

In the first situation the gradient will change according to the formula:

$$\tilde{g} = g + \alpha H p = g + \alpha \begin{pmatrix} R_V^T & 0 \\ S^T & R_F^T \end{pmatrix} \begin{pmatrix} w_V + \pi z_V \\ 0 \end{pmatrix}.$$

This means that

$$\tilde{g}_F = g_F + \alpha S^T (w_V + \pi z_V).$$

If a bound changes its status then the vector $\tilde{g}_F$ will change depending whether a bound has entered or left the working set.

If jth variable is added to the working set, a scalar γ is added at the front of $\tilde{g}_F$ to give:

$$\bar{g}_F = \begin{pmatrix} \gamma \\ \tilde{g}_F \end{pmatrix}.$$

The value of γ can be computed by multiplying the jth row of R_V^T by the vector $u_V + \alpha[w_V + \pi z_V]$ (but before any updates are done).

When a bound is deleted from the working set, the appropriate component of $\tilde{g}_F$ is removed and the remaining $n_F - 1$ components of $\tilde{g}_F$ form $\bar{g}_F$.

A.6 Modifications and Extensions

In the previous section we have assumed that the rows of the matrix A_V are linearly independent at each iteration. If this is not true our formulas (A.3.13)–(A.3.15), which we use as the basis for our algorithm, are no longer valid.

Assume that vectors a_i, $i \in I(x)$ defining active terms are linearly dependent, i.e.

$$a_k = \sum_{i \in I_1(x)} \beta_i a_i, \; I_1(x) \subset I(x), \; k \in I(x), \tag{A.6.1}$$

where $I_1(x)$ is the set of indices of the working set. If $\sum_{i \in I_1(x)} \beta_i \neq 1$ and $p \neq 0$ then

$$a_k^T p \neq a_i^T p = \pi \; \forall i \in I_1(x)$$

thus condition (A.3.7) required for the convergence of the *PLQP Algorithm* is not satisfied.

One way to rectify this problem is to transform the original problem (A.2.1)–(A.2.2) into the problem which has the same solution as problem (A.2.1)–(A.2.2), but is such that if (A.6.1) occurs than we always have $\sum_{i \in I_1(x)} \beta_i = 1$.

Consider the transformed problem (A.3.1)

$$\min_{\breve{x} \in \mathcal{R}^{n+1}} \breve{F}(\breve{x}), \tag{A.6.2}$$

where $\breve{F}(\breve{x})$ is given by

$$\breve{F}(\breve{x}) = \breve{c}^T \breve{x} + 1/2 \breve{x}^T \breve{H} \breve{x} + \max_{i \in I} \left[\breve{a}_i^T \breve{x} + b_i \right],$$

and $\breve{c}$, $\breve{a}_i$, $i \in I$, $\breve{H}$ are the enlarged vectors and the matrix of problem (A.3.1):

$$\breve{c} = \begin{pmatrix} c \\ 1 \end{pmatrix}, \quad \breve{H} = \begin{pmatrix} H & 0 \\ 0 & 1 \end{pmatrix}, \quad \breve{a}_i = \begin{pmatrix} a_i \\ -1 \end{pmatrix}. \tag{A.6.3}$$

A direction of descent for this problem, $\breve{p}$, has to satisfy the equations

$$\breve{H}\breve{p} + \breve{A}^T\lambda + \breve{g} = 0 \tag{A.6.4}$$
$$\breve{A}\breve{p} = \pi e \tag{A.6.5}$$
$$e^T\lambda = 1. \tag{A.6.6}$$

These equations can be reduced to equations (A.3.13)–(A.3.15) and one additional equation

$$(\breve{p})_{n+1} - e^T\lambda + (\breve{g})_{n+1} = 0. \tag{A.6.7}$$

Notice that if $(\breve{x}_{n+1}) = 0$ then $(\breve{g})_{n+1} = 1$, hence from (A.6.6)–(A.6.7) we obtain $(\breve{p})_{n+1} = 0$. If we start from a point $\breve{x}$ such that $(\breve{x})_{n+1} = 0$, then this property of $\breve{x}$ will be preserved during the course of the *PLQP Algorithm*. This means, from (A.6.3)–(A.6.6), that the first n components of the solution to problem (A.6.2) are the solutions of problem (A.3.1).

The advantage of this transformation is that the new problem is much less vulnerable to degeneracy. If $\breve{a}_k$ is linearly dependent on $\breve{a}_i$, $i \in I_1(x)$, i.e.,

$$\breve{a}_k = \sum_{i \in I_1(x)} \beta_i \breve{a}_i, \tag{A.6.8}$$

then $\sum_{i\in I_1(x)} \beta_i = 1$, from the last equation of (A.6.8), and finally $\breve{a}_k^T\breve{p} = \sum_{i\in I_1(x)} \beta_i \breve{a}_i^T\breve{p} = \sum_{i\in I_1(x)} \beta_i \pi = \pi$, as required. The extra computational burden associated with this transformation is not significant. Note that the extra variable does not have any bounds so it is always a varying variable. However, it is important to realize that this transformation does not remove all possible sources of *cycling* (c.f. [43]) in the *PLQP Algorithm*. If at the current point there are active (augmented) linear terms which are linearly dependent and we want to remove a linear term from the working set, then cycling can still occur. The standard perturbation technique (*relaxing* linear terms) can then be used to avoid it ([43]).

The *PLQP Algorithm* can also be used to solve the problems

$$\min_{x\in\mathcal{R}^n} \left[c^Tx + 1/2x^THx + r\max\left[\mu, \max_{i\in I}\left[a_i^Tx + b_i\right]\right]\right]$$
$$\text{s. t. } a_j^Tx + b_j \le \nu,\ j \in J,\ \ l \le x \le u \tag{A.6.9}$$

assuming that we know an initial point x which satisfies the constraints in the above problem.

The modified algorithm for problem (A.6.9) is very useful when repeatedly called to solve related problems in successive QP methods for constrained optimization ($\mu = \nu = 0$, $r > 0$ in this case). If H, a_i, b_i and c change but we estimate that the active set at the solution to problem (A.6.9) will not change then problem (A.6.9) with $I = I(x^l)$, $J = \emptyset$ and $\mu = 0$ is solved first. (Here x^l is the solution of (A.6.9) at the preceding call.) Next, its solution is used as an initial guess for the next problem (A.6.9) defined by the indices in the piecewise–linear term: $I \setminus I(x^l)$; and $J = I(x^l)$, $\mu = \nu = 0$. If at the solution to the first problem we have $\max_{i \in I(x^l)}[a_i^T x + b_i] > 0$, this means that the preceding $I(x^l)$ is not a good estimate for the active set at the solution for the current call. In this case we continue solving problem (A.6.9) with $J = \emptyset$ but with the full index set I. Otherwise ($\max_{i \in I(x^l)}[a_i^T x + b_i] \leq 0$), if at the solution to the second problem (A.6.9) we have $\max_{i \in I \setminus I(x^l)}[a_i^T x + b_i] > 0$, then the constraints $a_j^T x + b_j \leq 0$, $j \in I(x^l)$ are removed ($J = \emptyset$ is assumed) and solving problem (A.6.9) is continued with the full index set I. This strategy usually guarantees that the solution to the second problem is achieved in only few iterations. The extra burden associated with this approach results from the solution of equations (A.3.23) (because vector e has to be redefined to include some zeros, see (A.4.2)–(A.4.5)), and the calculation of z (due to (A.3.24)) when we switch from one problem to another. However, it must be mentioned that the corresponding procedure used in dual methods ([50], [65], [89]) is simpler and more straightforward.

If problem (A.6.9) is solved with several values of r (as in *Algorithm 4* described in Chapter 5), then simple modifications to the factorizations have to be made to continue calculations with a new value of r. Notice that when $r \neq 1$ the modification of c which resulted in $\breve{c} = [c, 1]$ should be substituted by $\breve{c} = [c, r]$ because instead of the relation $e^T\lambda = 1$ we have to work with $e^T\lambda = r$. This implies that u_V, v_L w_V g_V must be changed when r assumes a new value. First, the system of equations $R_V^T \hat{u}_V = \hat{g}_V$ has to be solved for $\hat{u}_V$, where $\hat{g}_V$ is the vector with all elements equal to zero except the one corresponding to the additional variable $\breve{p}_{n+1}$ which is equal to a new value of r (this can be done efficiently without solving the full set of equations, depending on the position of $\breve{p}_{n+1}$ in the vector $\breve{p}_V$). Next, the solution to these equations is used to update vectors u_V, w_V, v_L and the solution from the previous call is employed as the initial guess for the problem with the new r. This approach usually leads to very few iterations in the new problem. We are not aware of a similar simple procedure in dual methods. This is due to the fact that the term $c^T x$ is present in problem (A.6.9) thus rescaling primal, x, and dual, λ, variables simultaneously is not possible (see [65], [66]).

Finally, we would like to compare our method with other algorithms for problem (A.2.1)–(A.2.2). They are dual methods ([65],[112]) and that means that they maintain feasibility with respect to the Lagrange multipliers instead of the feasibility with respect to primal variables x as it is in our method. That dual strategy usually results in algorithms less vulnerable to cycling in the case of linearly dependent terms a_i. On the other hand dual methods do not cope so well with nearly singular matrices H because they start from the point $x = -H^{-1}c$ ([50])—according to Powell ([89]) this can be overcome to some extent. The methods described in [50], [89] and [65] are also based on the representation (A.3.16) of the matrix $R^{-T}A^T$ but the matrices $R^{-1}Q = (R^{-1}Y, R^{-1}Z) = (W, U)$, T are stored and updated instead of the matrices R, Y and T. This implies that these dual methods require extra memory to store the Cholesky factor R if it is to be updated in a master SQP algorithm. Furthermore, the box constraints are not treated in an efficient way in the mentioned papers. The modified Gram–Schmidt orthogonalization can also be applied in dual methods for problem (A.2.1)–(A.2.2), the example is given in [112] where the case with a diagonal Hessian matrix H is considered and box constraints are handled efficiently. Another dual approach to problem (A.2.1)–(A.2.2) is proposed in [66] where also only problems with diagonal Hessian matrices are analysed. The method in [66] pays special attention to box constraints.

But what really distinguishes our primal method from the dual methods is the way in which simple constraints (A.2.2) are tackled. If our method is used in successive QP algorithms, we can always initiate our method from the point which lies on the bounds active at the solution to the previously solved QP problem. This can effectively reduce the dimension of the problem and can improve conditioning of the matrix R since R_V can have significantly smaller dimension than R (see results for Example 1 in the next section). QP methods with this property are specially desirable in algorithms for optimal control problems whose solutions often lie on bounds. This initialization is not possible in dual methods because they always start from the feasible Lagrange multipliers which do not have to correspond to a vector x which satisfies simple constraints.

However, dual methods deserve more attention in the context of *FD Algorithm* introduced in Chapter 5 (esspecially when applied to problems described by large–scale differential–algebraic equations when CPU time devoted to QP problems is negligible in comparison to the total CPU required to solve the problems). The superlinear convergence properties of this method depends on the identification of the maximum set of linearly independent vectors (c.f. *Notation 5.4.1*) and that can require a QP procedure which copes well with *degenerate* QP problems. As mentioned above dual methods are much less upset by the degeneracy.

A.7 Numerical Experiments

In order to compare our procedure with other primal active set methods we implemented it and the subroutine LSSOL ([47]) in two algorithms for optimal control problems discussed in the monograph. LSSOL is a highly-regarded implementation of a null-space method that can accommodate a singular Hessian. PNTSOL ([96]) and LSSOL subroutines were implemented in *double precision* accuracy. The tests were performed on a Sun SPARCstation 10 Model 51.

The algorithms are based on the direction finding subproblem (**DFS**):

$$\min_{(d,\beta)} \left[\langle \nabla F(x^k), d\rangle + 1/2 d^T H^k d + r\beta\right] \tag{A.7.1}$$

subject to

$$\begin{aligned} \left|f_i^1(x^k) + \langle \nabla f_i^1(x^k), d\rangle\right| &\leq \beta \ \forall i \in E && \text{(A.7.2)}\\ f_j^2(x^k) + \langle \nabla f_j^2(x^k), d\rangle &\leq \beta \ \forall j \in I_\varepsilon(x^k) && \text{(A.7.3)}\\ l_d^1 \leq d &\leq u_d^1, && \text{(A.7.4)} \end{aligned}$$

where $I_\varepsilon(x^k)$ is a set of indices of the ε–active inequality constraints at the point x^k and H^k is a positive definite matrix; or the problems (**DFS′**):

$$\min_{(d,\beta)} \left[\|d\|^2 + \beta\right] \tag{A.7.5}$$

subject to

$$\langle \nabla F(x^k), d\rangle /r + \max_{i\in E} \left|f_i^1(x^k) + \langle \nabla f_i^1(x^k), d\rangle\right| - M(x^k) \leq \beta \tag{A.7.6}$$

$$f_j^2(x^k) + \langle \nabla f_j^2(x^k), d\rangle \leq \beta \quad \forall j \in I_\varepsilon(x^k) \tag{A.7.7}$$

$$l_d^1 \leq d \leq u_d^1, \tag{A.7.8}$$

where $M(x^k) = \max_{i\in E} |f_i^1(x^k)|$.

Moreover, if (**DFS**), or (**DFS′**), is solved to find a direction of descent, the second order correction direction is calculated by solving the problem (**SOC**):

$$\min_{d} \|d\|^2 \tag{A.7.9}$$

subject to

$$\begin{aligned} f_i^1(x^k + d^k) + \langle \nabla f_i^1(x^k), d\rangle &= 0 \ \forall i \in \tilde{I} && \text{(A.7.10)}\\ f_j^2(x^k + d^k) + \langle \nabla f_j^2(x^k), d\rangle &= 0 \ \forall j \in \tilde{E} && \text{(A.7.11)}\\ l_d^2 \leq d &\leq u_d^2, && \text{(A.7.12)} \end{aligned}$$

where $\tilde{I}$, $\tilde{E}$ are defined by the solutions to the problem (**DFS**), or the problem (**DFS′**).

The (**DFS**) are piecewise–linear quadratic problems (then β is superfluous), while the (**SOC**) can be solved by subroutine PNTSOL if an exact penalty function is applied to linear constraints (A.7.10)–(A.7.11) ([96]).

Three optimal control problems were solved by the two optimal control algorithms with different QP procedures for the direction finding subproblems. Therefore, PNTSOL and LSSOL solved similar (**PLQP**) problems generated by two master algorithms. We observed insignificant differences in data of the problems (**PLQP**) generated by the feasible directions algorithm with the different QP codes. The differences in the SQP algorithm were more noticeable due to slightly different values of the Lagrange multipliers associated with constraints (A.7.2)–(A.7.3) which were also used in building matrices H^k. However, these differences do not explain pronounced difference in the computational times recorded for the QP solvers.

The control problems are described in Chapter 4 (Examples 1–3). Table A.1 shows results for the feasible directions algorithm. In the second row a CPU time is reported for the LSSOL code when the discretization of control variables was $N = 100$. In the third row the results for PNTSOL are presented, while in the fourth one the results for PNTSOL but with the discretization $N = 1000$. (The corresponding time for LSSOL was so excessive that we do not present it here.) The numbers in brackets indicate percentage of time devoted to the direction finding subproblems in the optimal control algorithms. The second line in the top row describes the (**PLQP**) problems; the first number is the number of the (**DFS**) problems, the second number is the number of the (**SOC**) problems; while the last one is the average number of general linear constraints (overall all iterations) in the (**DFS**) or the (**DFS′**) problems.

When the algorithms were applied to control problems without state constraints (Problem 3) we performed calculations until solutions were found with the prespecified accuracy. Because a projection operation was used (in both algorithms) to cope efficiently with box constraints (c.f. Chapter 4) the number of iterations did not depend on the discretization parameter N. When the algorithms were tried on control problems with state constraints we fixed the number of iterations and state constraints were approximated as described in Chapter 4. Therefore the number of these constraints was dependent only on the discretization of state constraints which was the same for all runs for a given problem. The numbers of iterations were chosen to obtain good approximations to solutions for the smallest N considered. It must be mentioned that these numbers of iterations also provided controls close to solutions for other values of N in the case of the feasible directions algorithm. The efficient initialization to the problems (**DFS**) and (**DFS′**), as described in the previous section, was not applied in order not to distort the comparison of the PNTSOL and LSSOL codes. We think that the numerical results presented in the tables below give good indication of relative performance of null–space and range–space methods applied to the problems (**PLQP**) considered in this section.

Problem	1 31, 25, 24	2 72, 37, 25	3 28, 0, 5
LSSOL ($N = 100$)	24.42 (53.4%)	149.00 (50.8%)	0.97 (5.1%)
PNTSOL ($N = 100$)	1.91 (9.9%)	8.43 (9.2%)	0.18 (1.1%)
PNTSOL ($N = 1000$)	38.78 (13.9%)	370.08 (33.6%)	8.69 (4.7%)

TABLE A.1. Comparison of LSSOL and PNTSOL codes on the problems (**PLQP**) with diagonal Hessian matrices.

It is not surprising that the performance of PNTSOL is so superior in comparison with LSSOL. In the case of diagonal matrices H^k, because PNTSOL exploits this structure (but not LSSOL), the cost of one iteration in PNTSOL is proportional to the dimension N (if the number of piecewise–linear terms is fixed). Since the projection was used to cope with box constraints imposed on d these constraints active at the solution were typically identified after the first few calls to PNTSOL and that meant the box constraints were effectively eliminated from the problems (**DFS′**) and (**SOC**). This explains the moderate increase of a CPU time with respect to N. However, the computation time for $N = 1000$ is not excessive at all. The biggest is for the second problem, but in that case 109 problems were solved with 2000 variables and dense matrices of dimension 28×2000 corresponding to linear constraints (A.7.6)–(A.7.7). The average time for one problem was only 3.4 sec.

Table A.2 shows the corresponding results for the SQP algorithm. In this case the discretization $N = 1000$ has not been attempted because the matrices H^k generated by the algorithm were dense.

CPU times shown in Table A.2 are, in general, much higher than those stated in Table A.1. One cannot expect better characteristics of PNTSOL and LSSOL codes applied in the SQP algorithm since the matrices H^k are dense with N^2 elements. As before PNTSOL is much superior to LSSOL.

Problem	1 25, 25, 18	2 10, 10, 28	3 8, 0, 4
LSSOL $N = 100$ $N = 200$	 22.78 (51.9%) 90.54 (64.9%)	 29.51 (50.4%) 136.55 (55.0%)	 4.68 (42.5%) 40.59 (74.8%)
PNTSOL $N = 100$ $N = 200$	 4.25 (17.9%) 13.42 (25.9%)	 3.18 (10.8%) 10.56 (20%)	 0.25 (3.9%) 3.92 (25.5%)

TABLE A.2. Comparison of LSSOL and PNTSOL codes on the problems (**PLQP**) with dense Hessian matrices.

References

[1] Abadie, J and Carpentier, J. 1969 Generalization of the Wolfe Reduced Gradient Method to the Case of Nonlinear Constraints, in *Optimization*, R. Fletcher, ed., Academic Press, London, pp. 37–49.

[2] Alt, W. and Malanowski, K. 1995 The Lagrange–Newton Method for State-Constrained Optimal Control Problems. *Computational Optimization and Applications*, Vol. 4, pp. 217–239.

[3] Aubin, J.P. and Ekeland, I. 1984 *Applied Nonlinear Analysis*, Wiley-Interscience, New York.

[4] Bahri, P.A., Bandoni, J.A., and Romagnoli, J.A., 1996, Effect of Disturbances in Optimizing Control: Steady–State Open–Loop Backoff Problem. *AIChE J.*, Vol. 42, pp. 983–994.

[5] Baker, T.E. and Polak, E. 1994 On the Optimal Control of Systems Described by Evolution Equations, *SIAM J. Control and Optimization*, Vol. 32, pp. 224–260.

[6] Balakrishnan, A.V. and Neustadt, L.W. 1964 *Conference on Computing Methods in Optimization Problems*, editors, Academic Press, New York.

[7] Bartels, R.H., Golub, G.H. and Saunders, M.A. 1970 Numerical Techniques in Mathematical Programming, in *Nonlinear Programming*, J.B. Rosen, O.L. Mangasarian and K. Ritter, eds., Academic Press, New York, pp. 123–176.

[8] Barton, P.J. 1992 The Modelling and Simulation of Combined Discrete/Continuous Processes, *PhD Thesis*, University of London.

[9] Bellman, R. 1957 *Dynamic Programming*, Princeton University Press, Princeton, New Jersey.

[10] Bellman, R. and Dreyfus, S. 1962 *Applied Dynamic Programming*, Princeton University Press, Princeton, New Jersey.

[11] Berkovitz, L.D. 1974 *Optimal Control Theory*, Springer–Verlag, New York.

[12] Bertsekas, D.P. 1982 *Constrained Optimization and Lagrange Multiplier Methods*, Academic Press, New York.

[13] Bertsekas, D.P. 1995 *Dynamic Programming and Optimal Control*, Athena Scientific, Belmont, MA.

[14] Best, M.J. 1984 Equivalence of Some Quadratic Algorithms, *Mathematical Programming*, Vol. 30, pp. 71–87.

[15] Betts, J.T. 1992 Using Sparse Nonlinear Programming to Compute Low Thrust Orbit Transfers, *Beoing Computer Services Technical Report*, August 16.

[16] Betts, J.T. and Frank, P.D. 1994 A Sparse Nonlinear Optimization Algorithm, *J. Optimization Theory and Applications*, Vol. 82, pp. 519–541.

[17] Biegler, L.T., Nocedal, J. and Schmid, C. 1995 A Reduced Hessian Method for Large Scale Constrained Optimization, *SIAM J. Optimization*, Vol. 5, pp. 314–347.

[18] Boggs, P.T. and Tolle, J.W. 1994 Convergence Properties of a Class of Rank–Two Updates, *SIAM J. Optimization*, Vol. 4, pp. 262–287.

[19] Boggs, P.T. and Tolle, J.W. 1995 Sequential Quadratic Programming, *Acta Numerica*, Vol. 5, pp. 1–51.

[20] Breakwell, J.V. 1959 The Optimization of Trajectories, *SIAM Journal*, Vol. 7, pp. 215–247.

[21] Breakwell, J.V., Speyer, J. and Bryson, A.E. 1963 Optimization and Control of Nonlinear Systems Using the Second Variation, *SIAM J. on Control*, Vol. 1, pp. 193–223.

[22] Brenan, K.E., Campbell, S.L. and Petzold, L.R. 1989 *Numerical Solution of Initial–Value Problems in Differential–Algebraic Equations*, North–Holland, New York.

[23] Brooke, A., Kendrick, D. and Meeraus, M. 1988 *GAMS: A User's Guide*, Scientific Press, Redwood City.

[24] Bryson, A.E. 1996 Optimal Control—1950 to 1985, *IEEE Control Systems*, June, pp. 26–33.

[25] Bryson, A.E. and Denham, W.F. 1962 A Steepest Ascent Method for Solving Optimum Programming Problems, *ASME J. Appl. Mech.*, Vol. 29, Series E, pp. 247–257.

[26] Bryson, A.E. and Ho, Y.C. 1969 *Applied Optimal Control*, Blaisdell Publishing Company, Waltham MA.

[27] Bulirsch, R, Montrone, F. and Pesch, H.J. 1991 Abort Landing in the Presence of Windshear as a Minimax Optimal Control Problem, Part 1: Necessary Conditions. *J. Optimization Theory and Applications*, Vol. 70, pp. 1–23.

[28] Bulirsch, R, Montrone, F. and Pesch, H.J. 1991 Abort Landing in the Presence of Windshear as a Minimax Optimal Control Problem, Part 2: Multiple Shooting and Homotopy. *J. Optimization Theory and Applications*, Vol. 70, pp. 223–254.

[29] Burke, J.M. 1991 An Exact Penalization Viewpoint of Constrained Optimization, *SIAM J. Control and Optimization*, Vol. 29, pp. 968–998.

[30] Byrd, R. and Nocedal, J. 1991 An Analysis of Reduced Hessian Methods for Constrained Optimization, *Mathematical Programming*, Vol. 49, pp. 285–323.

[31] Chamberlain, R.M, Powell, M.J.D., Lemarechal, C. and Pedersen, H.C. 1982 The Watchdog Technique for Forcing Convergence in Algorithms for Constrained Optimization, *Mathematical Programming Study 16*, pp. 1–17.

[32] Clarke, F.H. 1990 *Optimization and Nonsmooth Analysis*, Classics in Applied Mathematics, SIAM Publications, Philadelphia.

[33] Colantonio, M.C. and Pytlak, R. 1997 Dynamic Optimization of Large–Scale Systems: Case Study, *International J. of Control*, Vol. 72, pp. 833–841.

[34] Coleman, T.F. and Liao, A. 1995 An efficient Trust Region Method for Unconstrained Discrete–Time Optimal Control Problems, *Computational Optimization and Applications*, Vol.6, pp. 47–66.

[35] Daniel, J.W., Gragg, W.B., Kaufman, L. and Stewart, G.W. 1976 Reorthogonalization and Stable Algorithms for Updating the Gram-Schmidt QR Factorization, *Mathematics of Computation*, Vol. 30, pp. 772–795.

[36] Dennis, J.E. and More, J.J. 1974 A Characterization of Superlinear Convergence and Its Application to Quasi–Newton Methods, *Mathematics of Computation*, Vol. 28, pp. 549–560.

[37] Dontchev, A.L. and Hager, W.W. 1998 Lipschitzian Stability for State Constrained Nonlinear Optimal Control, *SIAM J. Control and Optimization*, Vol. 36, pp. 698–718.

[38] Drud, A.S. 1985 A GRG Code for Large Sparse Dynamic Nonlinear Optimization Problems, *Mathematical Programming*, Vol. 31, pp. 153–191.

[39] Duff, I.S., Erisman, A.M. and Reid, J.K. 1989 *Direct Methods for Sparse Matrices*, Oxford Science Publications, Oxford.

[40] Dunn, J. and Bertsekas, D.P. 1989 Efficient Dynamic Programming Implementations of Newton's Method for Unconstrained Optimal Control Problems, *J. Optimization Theory and Applications*, Vol. 63, pp. 23–38.

[41] El-Alem, M. 1991 A Global Convergence Theory for the Celis–Dennis–Tapia Trust Region Algorithm for Constrained Optimization, *SIAM J. Numerical Analysis*, Vol. 28, pp. 266–290.

[42] Fedorenko, R.P. 1978 *Priblizhyonnoye reshenyie zadach optimalnovo upravlenya*, Nauka, Moskva.

[43] Fletcher, R. 1989 *Practical Optimization*, Chichester, J. Wiley & Sons.

[44] Fontecilla, R. 1988 Local Convergence of Secant Methods for Nonlinear Constrained Optimization, *SIAM J. Numerical Analysis*, Vol. 25, pp. 692–712.

[45] Gelfand, I.M. and Fomin, S.V. 1963 *Calculus of Variations*, Prentice–Hall, Englewood Cliffs.

[46] Gill, P.E., Gould, N.I.M., Murray, W., Saunders, M.A., and Wright, M.H. 1984 A Weighted Gram–Schmidt Method For Convex Quadratic Programming, *Mathematical Programming*, Vol. 30, pp. 176–195.

[47] Gill, P.E., Hammarling, S.J., Murray, W., Saunders, M.A. and Wright, M.H. 1986 User's Guide for LSSOL (Version 1.0), *Technical Report SOL 86-1*, Systems Optimization Laboratory, Department of Operations Research, Stanford University, California.

[48] Gill, P.E. and Murray, W. 1978 Numerically Stable Methods for Quadratic Programming, *Mathematical Programming*, Vol. 14, pp. 349–372.

[49] Girsanov, I.V. 1972 *Lectures on Mathematical Theory of Extremum Problems*, Springer–Verlag, New York.

[50] Goldfarb, D. and Idnani, A. 1983 A Numerically Stable Dual Method for Solving Strictly Convex Quadratic Programms, *Mathematical Programming*, Vol. 27, pp. 1–33.

[51] Goldstine, H.H. 1981 *A History of the Calculus of Variations from the 17th to the 19th Century*, Springer–Verlag, New York.

[52] Gramlich, G., Hettich, R. and Sachs, E.W. 1995 Local Convergence of SQP Methods in Semi–Infinite Programming, *SIAM J. Optimization*, Vol. 5, pp. 641–658.

[53] Hager, W.W. and Presler, D.L. 1987 Dual Techniques for Minmax, *SIAM J. Control and Optimization*, Vol. 25, pp. 660–685.

[54] Hairer, E., Lubich, Ch. and Roche, M. 1989 *The Numerical Solution of Differential-Algebraic Systems by Runge–Kutta Methods*, Lecture Notes in Mathematics 1409, Springer–Verlag, Berlin.

[55] Hairer, E. and Wanner, G. 1991 *Solving Ordinary Differential Equations, Vol.2 Stiff and Differential-Algebraic Problems*, Springer–Verlag, Berlin.

[56] Han, S.P. 1976 A Superlinearly Convergent Variable Metric Algorithms for General Nonlinear Programming Problems, *Mathematical Programming*, Vol. 11, pp. 263–282.

[57] Hartl, R.F., Sethi, S.P. and Vickson, R.G. 1995 A Survey of the Maximum Principles for Optimal Control Problems with State Constraints, *SIAM Review*, Vol. 37, pp. 181–218.

[58] *Harwell Subroutine Library Specification* 1993, pp. 261–273.

[59] Hauser, J.E. 1986 Proximity Algorithms: Theory and Implementation, *Memo. UCB/ERL M86/53*, Dep. Elect. Eng. Comput. Sci., Univ. California, Berkeley, CA.

[60] Hindmarsh, A.C. 1983 ODEPACK, a Systemized Collection of ODE Solvers, in *Scientific Computing*, R.S. Stepleman *et al* eds, North–Holland, Amsterdam, pp. 55–64.

[61] Jacobson, D. and Mayne, D.Q. 1970 *Differential Dynamic Programming*, Elsevier Science Publishing.

[62] Kantorovich, L.V. and Akilov, G.P. 1982 *Functional Analysis*, Pergamon Press, Oxford.

[63] Kingman, J.F.C. and Taylor, S.J. 1966 *Introduction to Measure and Probability*, Cambridge University Press, Cambridge.

[64] Kiwiel, K. 1985 *Methods of Descent for Nondifferentiable Optimization*, Springer–Verlag, Heidelberg.

[65] Kiwiel, K. 1989 A Dual Method for Certain Positive Semidefinite Quadratic Programming Problems, *SIAM J. Sci. Stat. Comput.*, Vol. 10, pp. 175–186.

[66] Kiwiel, K. 1994 A Cholesky Dual Method for Proximal Piecewise Linear Programming, *Numerische Mathematik*, Vol. 68, pp. 325–340.

[67] Kreindler, E. 1982 Additional Necessary Conditions for Optimal Control with State–Variable Inequality Constraints, *J. Optimization Theory and Optimization*, Vol. 38, pp. 241–250.

[68] Lawrence, C., Zhou, J.L. and Tits, A.L. 1994 User's Guide for CFSQP Version 2.1: A C Code for Solving (Large Scale) Constrained Nonlinear (Minimax) Optimization Problems, *TR-94-16rl*, Institute for Systems Research, University of Maryland.

[69] Logsdon, J.S. 1990 Efficient Determination of Optimal Control Profiles for Differential Algebraic Systems, *PhD Thesis*, Department of Chemical Engineering, Carnegie–Mellon University.

[70] Luenberger, D.G. 1973 *Introduction to Linear and Nonlinear Programming*, Addison–Wesley.

[71] Luenberger, D.G. 1998 *Investment Science*, Oxford University Press, New York.

[72] Machielsen, K.C.P. 1988 *Numerical Solution of Optimal Control Problems with State Constraints by Sequential Quadratic Programming in Function Space*, CWI Tract, Vol. 53, Amsterdam.

[73] Mayne, D.Q. and Polak, E. 1980 An Exact Penalty Function Algorithm for Optimal Control Problems with Control and Terminal Equality Constraints, Parts 1 and 2, *J. Optimization Theory and Applications*, Vol. 32, pp. 211–246, pp. 345–364.

[74] Mayne, D.Q. and Polak, E. 1982 A Superlinearly Convergent Algorithm for Constrained Optimization Problems, *Mathematical Programming Study*, Vol.16, pp. 45–61.

[75] Mayne, D.Q. and Polak, E. 1987 An Exact Penalty Function Algorithm for Control Problems with State and Control Constraints, *IEEE Trans. on AC*, Vol. AC–32, pp. 380–387.

[76] Miele, A., Mohanty, B.P., Vankataraman, P. and Kuo, Y.M. 1982 Numerical Solution of Minimax Problems of Optimal Control, Parts 1 and 2, *J. Optimization Theory and Applications*, Vol. 38, pp. 97–135.

[77] Miele, A. Wang, T, Tzeng, C.Y. and Melvin, W.W. 1987 Optimal Abort Landing Trajectories in the Presence of Windshear, *J. Optimization Theory and Applications*, Vol. 55, pp. 165–202.

[78] Mohideen, M.J., Perkins, J.D. and Pistikopoulos, E.N. 1995 Optimal Design of Dynamic Systems under Uncertainty, *AIChE Journal*, Vol. 42, pp. 2251–2272.

[79] Murray, W., and Wright, M.H. 1982 Computation of the Search Direction in Constrained Optimization Algorithms, *Mathematical Programming Study*, Vol. 16, pp. 62–83.

[80] Murtagh, B.A. and Saunders, M.A. 1982 A Projected Lagrangian Algorithm and its Implementation for Sparse Nonlinear Constraints, *Mathematical Programming Study*, Vol. 16, pp. 84–117.

[81] Oberle, H.J. 1986 Numerical Solution of Minimax Optimal Control Problems by Multiple Shooting Technique, *J. Optimization Theory and Applications*, Vol. 50, pp. 331–364.

[82] Ortega, J.M. and Rheinboldt, W.C. 1970 *Iterative Solution of Nonlinear Equations in Several Variables*, Academic Press, New York.

[83] Panier, E.R., and Tits, A.L. 1993 On Combining Feasibility, Descent and Superlinear Convergence in Inequality Constrained Optimization, *Mathematical Programming*, Vol. 59, pp. 261–276.

[84] Pantoja, J.F.A. De O. 1988 Differential Dynamic Programming and Newton's Method, *International J. Control*, Vol. 47, pp. 1539–1553.

[85] Polak, E. and He, L. 1991 Unified Steerable Phase I–Phase II Method of Feasible Directions for Semi–Infinite Optimization, *J. Optimization Theory and Applications*, Vol. 69, pp. 83–107.

[86] Polak, E., Yang, T.H. and Mayne, D.Q. 1993 A Method of Centers Based on Barrier Functions for Solving Optimal Control Problems with Continuum State and Control Constraints, *SIAM J. Control and Optimization*, Vol. 31, pp. 159–179.

[87] Pontryagin, L.S., Boltyanskii, V.G., Gamkrelidze, R.V. and Mischenko, E.F. 1962 *The Mathematical Theory of Optimal Processes*, Wiley, New York, New York.

[88] Powell, M.J.D. 1977 The Convergence of Variable Metric Methods for Nonlinearly Constrained Optimization Calculations, *Technical Memorandum No. 315*, Argonne National Laboratory, Argonne, Illinois.

[89] Powell, M.J.D. 1985 On the Quadratic Programming Algorithm of Goldfarb and Idnani. *Mathematical Programming Study*, Vol. 25, pp. 46–61.

[90] Pytlak, R. 1988 Strong Variation Algorithm for Minimax Control, *Lecture Notes on Control and Information Sciences*, Vol. 111, eds. A. Bensoussan, J.L. Lions, Springer–Verlag, Berlin, pp. 321–333.

[91] Pytlak, R. 1992 A Variational Approach to the Discrete Maximum Principle, *IMA Journal of Mathematical Control & Information*, Vol. 9, pp. 197–220.

[92] Pytlak, R. 1992 Numerical Methods for Optimal Control Problems with Input Control and State Inequality Constraints, *Proceed. of NOLCOS'92*, 24-26 June, 1992, Bordeaux, France, pp. 531–536.

[93] Pytlak, R. 1992 Numerical Experiment with a Method of Centers Based on Barrier Functions for Optimal Control Problems with State and Control Constraints, *Research Report PS92-64*, Centre for Process Systems Engineering, Imperial College, UK.

[94] Pytlak, R. 1993 PH2SOL: a Phase I–Phase II Solver for a General Optimal Control Problem with Hard State and Control Constraints, User's Manual, *Research Report C93-27*, Centre for Process Systems Engineering, Imperial College, UK.

[95] Pytlak, R. 1993 A Range–Space Method for Piecewise–Linear Quadratic Programming, *Research Report C93-13*, Centre for Process Systems Engineering, Imperial College, UK, also *Proceed. of the 33rd IEEE CDC*, Orlando, USA, 1994, pp. 1462-1463.

[96] Pytlak, R. 1993 PNTSOL: a Range–Space Solver for Piecewise–Linear Quadratic Programming, *Research Report C93-14*, Centre for Process Systems Engineering, Imperial College, UK.

[97] Pytlak, R. 1994 Optimal Control of Differential–Algebraic Equations, *Research Report C95-17*, Centre for Process systems Engineering, Imperial College, UK, also *Proceed. of the 33rd IEEE CDC*, Orlando, USA, pp. 951–956.

[98] Pytlak, R 1995 Second–Order Method for Optimal Control Problems with State Constraints, *Research Report C95-33*, Centre for Process Systems Engineering, Imperial College, UK, also *Proceed. of the 34th IEEE CDC*, St. Louis, USA, pp. 625–630.

[99] R. Pytlak 1995 SQPCON: an SQP Solver for a General Optimal Control Problem with State Constraints, *Research Report C95-34*, Centre for Process Systems Engineering.

[100] Pytlak, R. 1998 Runge–Kutta Based Procedure for Optimal Control of Differential–Algebraic Equations, *J. Optimization Theory and Applications*, Vol. 97, pp. 675–705.

[101] R. Pytlak 1997 RKCON: Runge–Kutta Based Procedure for Optimal Control of Differential–Algebraic Equations, User's Manual, *Research Report C97-22*, Centre for Process Systems Engineering, Imperial College, UK.

[102] Pytlak, R. 1998 An Efficient Procedure for Large–Scale Nonlinear Programming Problems with Simple Bounds on the Variables, *SIAM J. Optimization*, Vol. 8, No. 2, pp. 532–560.

[103] R. Pytlak 1997 An Algorithm for a Minimum Fuel Control Problem, *International J. of Control*, Vol. 72, pp. 435–448.

[104] Pytlak, R. and Malinowski, K. 1989 Optimal Scheduling of Reservoir Releases during Flood: Deterministic Optimization Problem, Part 1, Procedure, *J. Optimization Theory and Applications*, Vol. 61, pp. 409–432.

[105] Pytlak, R. and Malinowski, K. 1989 Optimal Scheduling of Reservoir Releases during Flood: Deterministic Optimization Problem, Part 2, Case Study, *J. Optimization Theory and Applications*, Vol. 61, pp. 433–449.

[106] Pytlak, R., Mohideen, M.J. and Pistikopoulos, E.N. 1996 Numerical Procedure for Optimal Control of Differential–Algebraic Equations, *Research Report C96-23*, Centre for Process Systems Engineering, Imperial College, UK.

[107] Pytlak, R. and R.B. Vinter 1993 PH2SOL Solver: an $O(N)$ Implementation of an Optimization Algorithm for a General Optimal Control Problem, *Research Report C93-36*, Centre for Process Systems Engineering, Imperial College, UK.

[108] Pytlak, R. and Vinter, R.B. 1996 A Feasible Directions Algorithm for Optimal Control Problems with State and Control Constraints: Convergence Analysis, *Research Report C96-24*, Centre for Process Systems Engineering, Imperial College, UK, also *SIAM J. Control and Optimization*, Vol. 36, pp. 1999–2019 (1998).

[109] Pytlak, R. and Vinter, R.B. 1999 A Feasible Directions Algorithm for Optimal Control Problems with State and Control Constraints: Implementation, *J. Optimization Theory and Applications*, Vol. 101, pp. 623–649.

[110] Rockafellar, R.T. 1970 *Convex Analysis*, Princeton University Press, Princeton.

[111] Rockafellar, R.T. and Wets, R.J. 1990 Generalized Linear–Quadratic Problems of Deterministic and Stochastic Optimal Control in Discrete Time, *SIAM J. Control and Optimization*, Vol. 28, pp. 810–822.

[112] Ruszczyński, A. 1986 A Regularized Decomposition Method for Minimizing a Sum of Polyhedral Functions, *Mathematical Programming*, Vol. 35, pp. 309–333.

[113] Shampine, L.F. 1980 Implementation of Implicit Formulas for the Solution of ODEs, *SIAM J. Sci. Stat. Comput.*, Vol. 1, pp. 103–118.

[114] SPEEDUP User Manual: Release 5.0 1988, Prosys Technology, Cambridge, UK.

[115] Stassinopoulos, G.I. and Vinter, R.B. 1978 Conditions for Convergence in the Computation of Optimal Controls, *J. Inst. Maths Applics*, Vol. 22, pp.1–14.

[116] Sussmann, H.J. and Willems, J.C. 1997 300 Years of Optimal Control: From the Brachystochrone to the Maximum Principle, *IEEE Control Systems*, July, pp. 32–44.

[117] Tanartkit, P. and Biegler, L.T. 1995 Stable Decomposition for Dynamic Optimization, *Ind. Eng. Chem. Res.*, Vol. 34, pp. 1253–1266.

[118] Teo, K.L. and Jennings, L.S. 1989 Nonlinear Optimal Control Problems with Continuous State Inequality Constraints, *J. Optimization Theory and Applications*, Vol. 63, pp. 1–22.

[119] Vassiliadis, V. 1993 Computational Solution of Dynamic Optimization Problems with General Differential–Algebraic Constraints, *PhD Thesis*, University of London, UK.

[120] Walsh, S., Chenery, S., Owen, P. and Malik, T. 1997 A Case Study in Control Structure Selection for a Chemical Reactor, *30th European Symposium on Computer Aided Process Engineering (ESCAPE–97), Trondheim, Norway,* Supplement to Computers and Chemical Engineering, ed. S. Skogestad, pp. S391–S396.

[121] Warga, J. 1972 *Optimal Control of Differential and Functional Equations*, Academic Press, New York.

[122] Warga, J. 1977 Steepest Descent with Relaxed Controls, *SIAM J. Control and Optimization*, Vol. 15, pp. 674–682.

[123] Warga, J. 1982 Iterative Procedure for Constrained and Unilateral Optimization Problems, *SIAM J. Control and Optimization*, Vol. 20, pp. 360–376.

[124] Wierzbicki, P. 1977 *Modele i wrazliwosc ukladow sterowania*, WNT, Warszawa (in Polish).

[125] Williamson, L.J. and Polak, E. 1976 Relaxed Controls and the Convergence of Optimal Control Algorithms, *SIAM J. Control and Optimization*, Vol. 14, pp. 737–757.

[126] Wright, S.J. 1991 Structured Interior Point Methods for Optimal Control, *Proceed. of the 30th CDC*, Brighton, UK, pp. 1711–1716.

[127] Wright, S.J. 1993 Interior Point Methods for Optimal Control of Discrete-Time Systems, *J. Optimization Theory and Applications*, Vol. 77, pp. 161–187.

[128] Young, L.C. 1969 *Lectures on the Calculus of Variations and Optimal Control Theory*, W.B. Saunders, Philadelphia.

[129] Zeidan, V. 1993 Second–Order Conditions for Optimal Control Problems with Mixed State–Control Constraints, *Proced. of the 32nd CDC IEEE*, San Antonio, pp. 3800–3805.

[130] Zhou, J.L. and Tits, A.L. 1996 An SQP Algorithm for Finely Discretized Continuous Minimax Problems and Other Minimax Problems with Many Objective Functions, *SIAM J. Optimization*, Vol. 6, pp. 461–487.

[131] Zhu, C. 1995 On the Optimal Primal–Dual Descent Algorithm for Extended Linear–Quadratic Programming, *SIAM J. Optimization*, Vol. 5, pp. 114–128.

[132] Zhu, C. and Rockafellar, R.T. 1993 Primal–Dual Projected Algorithms for Extended Linear–Quadratic Programming, *SIAM J. Optimization*, Vol. 3, pp. 751–783.

List of Symbols

Frequently used symbols:

$\mathcal{R}^n$	n–dimensional Euclidean space
$\|\cdot\|_{\mathcal{R}^n}$	Euclidean norm of n–dimensional space, but also $\|\cdot\|$ when it is convenient
$\langle\cdot,\cdot\rangle_{\mathcal{R}^n}$	Euclidean scalar product of n–dimensional space, but also $\langle\cdot,\cdot\rangle$ when it is convenient
$\mathcal{L}^p_m[T]$	$(1 \le p < \infty)$ space of measurable functions x such that $x(t) \in \mathcal{R}^m$ and $\int_T \|x(t)\|^p dt < \infty$
$\|\cdot\|_{\mathcal{L}^p}$	norm in $\mathcal{L}^p_m[T]$
$\mathcal{L}^\infty_m[T]$	space of measurable functions x such that $x(t) \in \mathcal{R}^m$ and $\operatorname{ess\,sup}_{t\in T} \|x(t)\| < \infty$
$\|\cdot\|_{\mathcal{L}^\infty}$	norm in $\mathcal{L}^\infty_m[T]$
$\mathcal{C}(T,\mathcal{R}^n)$	space of continuous n valued functions on compact T. If $n = 1$ we write $\mathcal{C}(T)$ in place of $\mathcal{C}(T,\mathcal{R})$.

Other symbols:

$\mathcal{U}$	13 (Chapter 2), 28 (Chapter 3)
x^u	13 (Chapter 2)
$y^{u,v}$	13 (Chapter 2)
$R_{\varepsilon,u}$	17 (Chapter 2)
$\langle\cdot,\cdot\rangle_{\mathcal{X}}$	20 (Chapter 2)
$A^\star$	20 (Chapter 2)
$L(x,u,p)$	22 (Chapter 2)
$H(t,x(t),u(t),p(t))$	22 (Chapter 2)
Ω	27 (Chapter 3)
$\tilde{F}_0$, $\tilde{h}^1_i$, etc.	28 (Chapter 3)
$\tilde{Q}$	28 (Chapter 3)
$d\tilde{Q}(u,h)$	28 (Chapter 3)
$\tilde{F}_c$	29 (Chapter 3), 83 (Chapter 4)
$\langle\nabla\tilde{F}_0(u),d\rangle$, $\langle\nabla\tilde{h}^1_i(u),d\rangle$, etc.	31 (Chapter 3)
$\sigma_c(u)$	34 (Chapter 3)
$\mathcal{L}^1(T,\mathcal{C}(\Omega))$	32 (Chapter 3)
$\hat{F}_0(\mu)$, $\hat{h}^1_i(\mu)$, etc.	32 (Chapter 3)
$\mathcal{D}$	33 (Chapter 3)
$\langle\nabla\hat{F}_0,d\rangle_r$, $\langle\nabla\hat{h}^1_i(\mu),d\rangle_r$, etc.	33 (Chapter 3)

$\mathcal{E}(\mu)$, $\mathcal{F}(\mu)$	36 (Chapter 3)
$P_{\mathcal{U}}[\cdot]$	57 (Chapter 4
$\mathcal{U}_{\omega,u}$	57 (Chapter 4)
$\sigma_{c,A}(u)$	58 (Chapter 4)
$t_{c,A}(u)$	58 (Chapter 4)
$\mathcal{L}_m^N[T]$	67 (Chapter 4)
$\vec{u}^N$, $\vec{u}$	67 (Chapter 4)
$\nabla\bar{F}_0(\vec{u})$, $\nabla\bar{h}_i^1(\vec{u})$, etc.	67 (Chapter 4)
$\vec{\mathcal{U}}$	68 (Chapter 4)
$\bar{F}_0(\vec{u})$, $\bar{h}_i^1(\vec{u})$, etc.	68 (Chapter 4)
$\sigma_c^H(u)$	84 (Chapter 5)
$t_c^H(u)$	84 (Chapter 5)
$\sigma_{c,A}^H(u)$	87 (Chapter 5)
$t_{c,A}^H(u)$	87 (Chapter 5)
$\mathcal{E}(\vec{u})$, $\mathcal{F}(\vec{u})$	88 (Chapter 5)
$\bar{h}_{i,1}^1$, $\bar{h}_{i,2}^1$	94 (Chapter 5)
$E^+(\vec{u})$, $E^-(\vec{u})$, $T(\vec{u})$, $I(\vec{u})$	95 (Chapter 5)
$E(\vec{u})$	98 (Chapter 5)
$\mathcal{B}(\vec{u},\delta)$	100 (Chapter 5)
t^m, $m^t(\vec{u})(\cdot)$	107 (Chapter 5)
$\mathcal{M}(\vec{u})$	107 (Chapter 5)
$T^c(\vec{u})$, $T^v(\vec{u})$	108 (Chapter 5)
$\mathcal{K}(\vec{u})$, $\mathcal{K}^s(\vec{u})$	108 (Chapter 5)
$L_k^q(t)$, $L_k^{h^2}(j)$, $L_k^{h^1}(i)$	113 (Chapter 5)
$\hat{T}(\vec{u})$	115 (Chapter 5)
$J^s(k)$	139 (Chapter 6)
$\sigma_{c,A,H}^{DAE}(u)$	142 (Chapter 6)
$t_{c,A,H}^{DAE}$	142 (Chapter 6)
$R_{\varepsilon,\vec{u}}^d$	141 (Chapter 6)

Optimization problems:

$(\mathbf{P})$	27 (Chapter 3)
$(\mathbf{P^r})$	33 (Chapter 3)
$\mathbf{P_c(u)}$	34 (Chapter 3)
$\mathbf{P_{c,A}(u)}$	57 (Chapter 4)
$(\mathbf{P^N})$	68, 82 (Chapter 4)
$(\mathbf{P_{NLP}(\vec{u})})$	82 (Chapter 5)
$(\mathbf{P_c})$	83 (Chapter 5)
$\mathbf{P_c^H(u)}$	84 (Chapter 5)
$\mathbf{P_{c,A}^H(\vec{u})}$	87 (Chapter 5)
$\mathbf{P_{A,H}^{SQP}(\vec{u})}$	94 (Chapter 5)
$(\mathbf{P_{DAE}})$	129 (Chapter 6)
$(\mathbf{P_{DAE}^N})$	140 (Chapter 6)
$\mathbf{P_{c,A,H}^{DAE}(\vec{u})}$	142 (Chapter 6)

(PLQP)	170 (Appendix)
$\mathbf{P}(x, J)$	172 (Appendix)
(DFS)	192 (Appendix)
(DFS')	192 (Appendix)
(SOC)	192 (Appendix)

Hypotheses:

(H1)	13 (Chapter 2)
(H2)	36 (Chapter 3)
(CQ)	36 (Chapter 3)
(H3)	56 (Chapter 4)
(BH)	84 (Chapter 5)
$(\mathbf{BH}^{\mathrm{N}})$	87 (Chapter 5)
$(\mathbf{CQ}^{\mathrm{N}})$	88 (Chapter 5)
(H4)	98 (Chapter 5)
(H5)	101 (Chapter 5)
(H6)	107 (Chapter 5)

Subject Index

accessory problem, 8, 81, 167
active arc, 8, 92
active set method, 69, 170
adjoint equations, 20
    continuous time, 22, 67
        for implicit systems, 132
    discrete time, 25
        for implicit systems, 138
adjoint operator, 20
    self–adjoint operator, 84
algebraic state, 129
approximation errors, 9
Ascoli's theorem, 15

backoff problem, 161
backward differentiation formula (BDF), 133
barrier function algorithm, 29
barycentric coordinates, 42
basic existence/uniqnuess theorem, 14
Bellman's dynamic programming equation, 6
BFGS updating formula, 125
BLAS (Basic Linear Algebra Subroutines), 68
brachistochrone problem, 2, 9, 69

calculus of variations, 1
Chebyshev functional, 28
Cholesky factorization, 173
collocation method, 8, 83
constraint qualification, 36, 88
control function, 1, 13
control system, 13
cycling, 189

descent function, nonpositive 34, 58, 84, 87, 142
differential–algebraic equations
    implicit, 129
    semi–explicit, 136
    index one, 130, 136
differential state, 129
Dini derivative, 28
direction finding subproblem, 34
discrete time system, 137
    implicit, 137
dominated convergence theorem, 15
dual method, 171, 190

endpoint constraints, 29
Euler–Lagrange equation, 2, 3
exact penalty function, 29, 83, 142

gradient algorithms, 7

feasible directions algorithm, 27

gradient restoration algorithm, 76
Gauss integration procedure, 136
Givens matrix, 179
Gram–Schmidt factorization, 176
Gronwall's lemma, 14, 60

Hamilton–Jacobi equation, 5, 6

Hamiltonian, 22
Hessenberg matrix, 178
Hessian, 20
 reduced, 21

implicit function theorem, 20

Lagrange multiplier, 98, 172
Lagrangian, 20
linearized equations, 13
$\mathcal{L}$–stable integration procedure, 137

maximum principle
 Pontryagin's, 6
 weak, 63, 39, 60
  discrete time, 140
maximum set of linearly independent vectors, 98
mean value theorem, 16
minimax theorem, 52
minimum time problem, 2, 69
multiple shooting method, 76
multipoint boundary value problem, 76

Newton step, 65
nonlinear programming problem (NLP), 82
null–space method, 170

optimality conditions
 necessary, normal, 38
 sufficiency, 98

pathwise inequality constraint, 6, 27
piecewise constant approximation, 24
piecewise–linear quadratic programming problem, 170
penalty test function, 34, 58, 84, 87, 142
plane rotation, 178
projection, 57
 matrix, 105
proximity algorithm, 58
point of attraction, 101
Powell–symmetric–Broyden update (PSB), 121
primal method, 171

Radau IIA integration procedure, 136, 144
Radon probability measure, 32
range–space method, 69, 170
relaxed controls, 32
relaxed dynamics, 32
relaxed necessary optimality conditions, 38
relaxed problem, 33
relaxed state trajectory, 32
reorthogonalization, 179
Runge–Kutta integration procedure
 explicit, 134
 implicit, 129, 134

search directions, set, 33
second order correction, 65
semi–infinite programming (SIP), 67, 82
sensitivity equations, 131
sequential quadratic programming (SQP), 86
 reduced gradient, 130
simplex, 42
singular subarc, 77
state trajectory, 1, 13
stepsize selection procedure, 146
stiff equations, 147
strict complementarity condition, 99
strong local minimum, 99
subgradient, 65
superlinear convergence, 104, 82

two–step, 122

uniform approximation, $(R_{\varepsilon,\mu}, \xi)$,
56

watchdog technique, 107, 110
windshear problem, 74

Printing: Weihert-Druck GmbH, Darmstadt
Binding: Buchbinderei Schäffer, Grünstadt

4. Lecture Notes are printed by photo-offset from the master-copy delivered in camera-ready form by the authors. Springer-Verlag provides technical instructions for the preparation of manuscripts. Macro packages in TEX, LATEX2e, LATEX2.09 are available from Springer's web-pages at http://www.springer.de/math/authors. Careful preparation of the manuscripts will help keep production time short and ensure satisfactory appearance of the finished book.

The actual production of a Lecture Notes volume takes approximately 12 weeks.

5. Authors receive a total of 50 free copies of their volume, but no royalties. They are entitled to a discount of 33.3 % on the price of Springer books purchase for their personal use, if ordering directly from Springer-Verlag.

Commitment to publish is made by letter of intent rather than by signing a formal contract. Springer-Verlag secures the copyright for each volume. Authors are free to reuse material contained in their LNM volumes in later publications: A brief written (or e-mail) request for formal permission is sufficient.

Addresses:

Professor F. Takens, Mathematisch Instituut,
Rijksuniversiteit Groningen, Postbus 800,
9700 AV Groningen, The Netherlands
E-mail: F.Takens@math.rug.nl

Professor B. Teissier, DMI, École Normale Supérieure
45, rue d'Ulm,
F-7500 Paris, France
E-mail: Teissier@ens.fr

Springer-Verlag, Mathematics Editorial, Tiergartenstr. 17,
D-69121 Heidelberg, Germany,
Tel.: *49 (6221) 487-701
Fax: *49 (6221) 487-355
E-mail: C.Byrne@Springer.de

9 783540 662143